ANIMALS
OF
AUSTRALIA

1. Kangaroo. Pencil sketch. Parkinson, 1770. *British Museum (Natural History).*

AXEL POIGNANT

ANIMALS
OF
AUSTRALIA

DODD, MEAD AND COMPANY
NEW YORK

This edition published in 1967 by

DODD, MEAD AND COMPANY, INC.

79 MADISON AVENUE

NEW YORK 10016

First published in 1965 by
ANGUS AND ROBERTSON LTD
LONDON, SYDNEY AND MELBOURNE
under the title
THE IMPROBABLE KANGAROO and other Australian Animals

PRINTED IN THE NETHERLANDS
BY L. VAN LEER & CO. N.V.
PHOTOTYPESET BY
OLIVER BURRIDGE FILMSETTING LIMITED
ENGLAND

Author's Note

PHOTOGRAPHY is only a means to an end, not the end itself. In my case it has helped my own interest in Australian animals to grow to a stage where I discovered for myself every animal's essential right to its own life. When photographing animals one is often privileged to witness moments in their private lives that reveal their skilful adaptation to situations and circumstances. This made me realize that humans have no proprietary right to the world and have no mandate to destroy or change the environment without the greatest care and consideration. I am not a naturalist, and undoubtedly better natural history photographs than these have been taken; but these pictures are the fruit of my own discovery of Australian animals. I only wish the results measured up to the respect I feel for the subject.

When assembling these photographs I began to think how short a time it was since Australian animals became known to the rest of the world—less than 200 years—and I became interested in the reactions people in Europe must have had on first learning about them. My wife, Roslyn, having a background of research into Australian history, began to cull the accounts of the early explorers and settlers for their comments and sought out the early drawings that they made. This book is the result.

My thanks are due to the countless friends who have helped me get my photographs and encouraged Roslyn and me with this book, to the naturalist Vincent Serventy who, many years ago, showed me the way, and also to Eric Worrell, who taught me a proper regard for snakes. I am also grateful to Mr Frank Shepherd of the Shepherd Press for allowing me to reproduce here the photographs numbered 51, 53, 87, and 184, from *Bush Animals of Australia*, published by him. A special salute to M. A. Maury, Director of the Le Havre Museum, France, who during the last war courageously saved Lesueur's Australian drawings and papers from both the Nazis and the British bombers. I thank him for allowing me to reproduce some of the paintings here. George Stubbs's paintings "A Portrait of the Kongouro" and "A Large Dog" are included by kind permission of Mrs W. P. Keith.

A great deal of natural history material collected during the early years of Australian exploration and settlement is now stored in the British Museum (Natural History). In drawing on this we have been shown much courtesy and help which I gratefully acknowledge. Perhaps my greatest thanks are due to Roslyn, who has worked so hard to reduce the wealth of historical information to manageable proportions and edit the notes into something much better than I could ever have done.

AXEL POIGNANT

Contents and Index

IN THIS INDEX the numbers refer to the illustrations rather than the pages, except that reference is also made to the text immediately adjacent to the pictures.

17th APRIL 1770: "A Gannet was seen which flew towards the NW with a steady uninterrupted flight as if he knew the road he was going to lead to the shore." This was the entry Joseph Banks made in his journal aboard the *Endeavour*. For several days the men had been reading nature's signs that land was at hand: a butterfly, an off-course land-bird, shearwaters and drifting seaweed. Two days later they sighted the eastern coast of New Holland, as Australia was then called, and turned northward. They found no sophisticated communities, no precious metals or spices or other useful items of trade, but they were not repelled by the country as earlier navigators had been on the western coast. Exploration was soon to be followed by settlement.

The gannet's flight heralded the end of the isolation of the Australian continent which had been brought about by its separation from the Asian land mass some fifty million years before.

The Aborigines, who did not reach Australia until, perhaps, about twenty thousand years ago, were nomadic people who neither grazed animals nor grew crops. Except for the fires they lit and the wild dog that came with them, their presence did little to disrupt the environment.

In this isolation, free from competition from the higher mammals, the marsupial population evolved a variety of forms to suit different habitats of desert, plain, forest and scrub. In the process of adaptation some acquired superficial resemblances to the non-marsupial mammals of other parts of the world. Therefore, it is not surprising that the Europeans who saw them first were both puzzled by the suggestion of animals already familiar to them and startled by their uniqueness.

The earliest accounts come from Dutch and English navigators of the seventeenth century who visited the west coast and returned with tales of a barren country and strange beasts. Both Pelsaert and Dampier described animals that hopped. Pelsaert called those he found on the Abrolhos Islands in 1629 "a species of cats, which are very strange creatures", and he observed: "Below the belly the female carries a pouch." Such descriptions gained little more credence than the story Dampier told of a preposterous creature with a "Head at each End. . . ." The contrary nature of things from the antipodes was confirmed when another Dutchman, de Vlamingh, arrived in Batavia in 1697 with three swans, which were black. These were the first representatives of Australian wildlife to be exported.

By the time H.M.S. *Endeavour* sailed from England in 1768, under the command of Lieutenant James Cook, to carry out astronomical observations in the Pacific and to search for the Great South Land, exploration had acquired a new scientific emphasis. Aboard the *Endeavour* was a party of naturalists and artists, led by Joseph Banks and equipped by him to the extent of some £10,000 of his own money, to observe, make records and collect specimens. Early in the voyage Banks contemplated the prospect of making new discoveries, and he wrote in his journal: "Dr Solander and my self shall have probably greater opportunity in the course of this voyage than any one has had before us. . . ." And he hoped that what was found would "add considerable Light to the science which we so eagerly Pursue."

Perhaps their greatest opportunities came when, almost two years later, they reached the eastern shores of New Holland. Banks continued to record and speculate in his journal with an impressive freshness and liveliness. Everything interested him. He wrote about "a small kind of Caterpiler, green and beset with many hairs" which was a "wrathfull militia" amongst the mangroves, and about clouds of butterflies: ". . . the eye could not be turnd in any direction without seeing millions. . . ." He remarked also that "a Crow in England tho in general sufficiently wary is I must say a fool to a New Holland crow and the same may be said of almost if not all the Birds in the countrey."

Most interesting of all, it is possible, from the journals of both Cook and Banks, to piece together the story of the finding of the kangaroo. The first sign was at Botany Bay where they saw "the dung of a large animal that had fed on grass." Not until they spent seven weeks repairing the ship at the Endeavour River, where Cooktown now stands, were they able to examine the country properly.

There, on 23rd June, Cook wrote that "one of the men saw an animal something less than a greyhound, it was of a Mouse Colour very slender made and swift of foot." On the following

2. "A Portrait of the Kongouro from New Holland".
Stubbs, 1771-2.   *Mrs W. P. Keith, London.*

day he saw it himself and noted that it had "a long tail which it carried like a greyhound, in short I should have taken it for a wild dog, but for its walking or running in which it jumped like a Hare or a dear." Others described it has having a footprint like a goat, but Cook was doubtful, because "the length of the grass hindered my seeing its legs." Banks, on the same day, wrote, doubtless with some frustration, "Gathering plants and hearing descriptions of the animal which is now seen by every body"—except himself. The next day, 25th June, he did see it, "tho but imperfectly," and, unlike the others who sought to compare it with animals known to them, he wrote: "What to liken him to I could not tell, nothing certainly that I have seen at all resembles him."

On 7th July a hunt was organized. His greyhound failed to catch any of the animals: ". . . they beat him owing to the length and thickness of the grass which prevented him from running while they at every bound leapd over the tops of it. We observd much to our surprise that instead of Going upon all fours this animal went only upon two legs, making vast bounds. . . ." Seven days later a kangaroo was shot and in all three were taken. Banks summed up: "It is different from any European and indeed any animal I have heard or read of except the Gerbua of Egypt, which is not larger than a rat when this is as large as a midling lamb; the largest we shot was 84 lb. It may however be easily known from all other animals by the singular property of running or rather hopping upon only its hinder legs carrying its fore bent close to its breast. . . ." He wrote that the natives of the Endeavour River called it *Kangooroo*.

Sydney Parkinson, one of the artists of the expedition, made two pencil sketches of the Kangaroo (1) which remained unfinished because he died on the voyage home. Though the explorers saw this animal more frequently than any other, they still did not discover that the female had a pouch.

The great collection taken home by Banks and Solander was described in a letter to Linnaeus as "the greatest treasure of Natural History that ever was brought into any country at one time by two persons." Some of it was made known in the eagerly awaited official account of the voyage published by Hawkesworth in 1773. Much additional material found its way into other specialized publications during the years that followed. Banks corresponded widely about the discoveries, for instance, with the great naturalist de Buffon, who drew on them for his *Histoire Naturelle*. An engraving of the Blue-bellied Parrot of Botany Bay done in 1774 appeared in Peter Brown's *Nouvelle Illustrations de Zoologie*. Both the white and black cockatoos and the bustard were included in John Latham's *A General Synopsis of Birds* (1787). The animals found by the explorers—the kangaroo, the opossum (ring-tailed possum), the Je-Quoll (native cat) and the wild dog (dingo)—were discussed by Thomas Pennant in his *History of Quadrupeds* (1781).

Of all the material brought back to Europe nothing did more to intrigue society than the accounts of the improbable kangaroo. Almost immediately Banks commissioned George Stubbs to paint a picture of the animal (2). Stubbs probably inflated one of the skins brought

3. Kangaroo. Engraving from J. Hawkesworth's official account of the *Endeavour* voyage, 1773.

back and used it as a model. In 1773 the picture was exhibited by the Society of Artists. The Hawkesworth account of the same year contained an engraving of a kangaroo (3). As time passed the Stubbs painting was forgotten and it was generally assumed that the Hawkesworth engraving was derived from Parkinson's pencil sketches.

Then, in 1957, the painting was included in an exhibition of Stubbs's work at the Whitechapel Art Gallery, London, and a naturalist, Dr Averil Lysaght, pointed out that the Hawkesworth engraving was a mirror image of the painting. Europe's first understanding of the appearance of the kangaroo came therefore from Stubbs's vision of the animal transmitted through the medium of the Hawkesworth engraving. The kangaroo was well served by Stubbs both in the accuracy of his observation and in his feeling for the subject. Some of this was already lost in the engraving. Much more was lost in the many later illustrations which in their turn derived from this and bear little resemblance to the original, except for the stance of the animal (4, 5).

Naturalists speculated about the kangaroo's relationship to the other animals. Some classed it with the jerboa because of its manner of hopping. Others grouped it with the opossum because of its teeth. The German edition of Buffon's *Histoire Naturelle* (1783) went so far as to suggest that "perhaps the future will discover what until now we can only guess that the female . . . is equipped with a pouch."

The next opportunity to find out more about the animals of New South Wales came in 1788 when Captain Phillip established the first settlement at Port Jackson. As the first fleet sailed into Sydney Cove it seemed to Arthur Bowes, assistant surgeon on the *Lady Penrhyn*, that: "The singing of the various birds amongst the trees, and the flight of the numerous parraquets, lorrequets, cockatoos and maccaws made all around appear like enchantment." This exotic image was not to prevail. The first years of settlement were times of grim hardship: when more and more convicts, starved, diseased and often dying, poured into the colony. Nothing seemed to grow, the imported livestock strayed or died, and the natural products of the country provided little in the way of nourishment. Implements and clothing were scarce. Fear and loneliness, aroused by the all-surrounding bush, was almost overwhelming.

Yet amateur naturalists abounded. The officers and men, convicts and settlers, made drawings, wrote descriptions and collected specimens. From the very beginning every ship returning to England took back this material and also live birds and animals. This intense interest in natural history is exemplified by the way in which Surgeon-General White digressed from his account of the murder of two convict rush-cutters by the aborigines to give a lengthy description of the "yellow-eared flycatcher." Undoubtedly the need to live off the land brought many of the animals to the notice of the settlers. The first emu was shot in February 1788 and was said to taste not unlike young tender beef. The kangaroo, which supplied most of the fresh meat of the settlement, was thought to be like veal when young, though tough and stringy when old—"in Europe it would not be reckoned a delicacy."

4. Kangaroo. Engraving from *The Voyage of Governor Phillip to Botany Bay*, 1789. Underlines the difficulty European artists had in comprehending what this animal was really like.

Parrots, bustards and swans all went into the stewpot. Later some of the first wombats, koalas and lyrebirds to be discovered were eaten.

The settlers soon noticed that the kangaroo carried and nourished its young in a pouch; even more surprising, many other animals (whether they burrowed, hopped, climbed or glided) seemed to share the same characteristic. Seven of these pouched animals were discussed and illustrated in the first account of the colony, *The Voyage of Governor Phillip to Botany Bay*, published by J. Stockdale in 1789. The editorial comment declared, however, that what had been recorded in the eleven months of settlement hardly compared with the value of Banks's and Solander's observations made during their brief visit, and proclaimed "the great advantage of a scientific eye over that of the unlearned observer."

When, a few months later, John White's Journal of a *Voyage to New South Wales* was published, it included detailed descriptions of some of the animals by the anatomist John Hunter, who had examined preserved specimens. It also outlined the difficulties of classifying these new animals: ". . . they are, upon the whole, like no other that we yet know of; but as they have parts in some respect similar to others, names will naturally be given to them expressive of those similarities . . . for instance, one is called the Kangaroo Rat . . . which should not be called either Kangaroo or Rat. . . ." To solve this dilemma it was suggested that "the name given by the natives" should be collected—an admirable idea that was seldom followed. *Koala, wombat* and *kangaroo* derive from aboriginal words, but the common names of most marsupials continued to be inspired by their resemblance to mammals of other lands.

Discoveries of interest to naturalists continued to be featured in the various accounts by other officers in the colony published during the 1790s. Much of the material also made its way via the Governors' dispatches and private correspondence. Almost as remarkable as the quantity of material collected was the speed with which it was disseminated amongst the scientists and the general public in Europe.

The animals were the subjects of papers delivered to the Royal Society and the Linnean Society. The first specimen of a platypus went to the Literary and Philosophical Society of Newcastle upon Tyne. The English ornithologist John Latham and the naturalist Thomas Pennant included the animals in new editions of their books; new books, like George Shaw's *Zoology of New Holland* (1794), were devoted entirely to the subject.

Some of the illustrations for these books were done in England, but many were derived from original drawings done in the colony. In fact, the copying of drawings and their distribution amongst the naturalists demonstrate the eagerness there was for knowledge about the animals. A valuable discussion about the implications of this, and about the identity and work of these early artists, is to be found in Bernard Smith's *European Vision and the South Pacific*.

Dr Smith gives the name "Port Jackson Painter" to one of the most skilled of these early anonymous artists (89). Some of his work and that of other anonymous artists is grouped in a collection known as the Watling Drawings, which also includes drawings signed by Thomas

5. Kangaroos with a "Hawkesworth ancestry". Possibly it is Mr Banks in the background. Engraving by T. Edwards. *Museum Leverianum*, 1795.

Watling (185). Watling was transported for forgery and, for several years after arriving in the colony in 1792, applied his skill to drawing for John White.

From all these sources there emerges a picture of tremendous interest in the natural history of the new settlement. The youthful naturalist of the *Endeavour* became the eminent and influential Sir Joseph Banks, President of the Royal Society from 1778, and he never lost interest in the young colony. His house in Soho Square became both a repository and a clearing station for all items of interest from New South Wales. His correspondence is a mine of information. The scientist Broussonet wrote thanking Banks for the kangaroos which had arrived in France as early as 1789. Gifts of the animals from New South Wales became a way of exchanging courtesies. A platypus was given to Napoleon in 1802, but it would seem from the exchange of letters between Sir Charles Blagden and Sir Joseph Banks that a pair of live kangaroos would have been considered more acceptable. For, wrote Sir Charles: "The park of St Cloud where he is going to reside, would be an excellent place for a little paddock of Kangaroos, like that of the King at Richmond." When, several months later, two kangaroos did arrive, they came not as a gift from the English king but from the animal dealer Pidcock, and Blagden lamented: "I am sorry for it, as for an excellent opportunity missed."

Australia's animals rated high as the favoured occupants of the parks of the great country houses. A dingo was kept at Hatfield House as early as 1789. There were possums in Durham, great gliders in Oxfordshire, and the Earl of Exeter had emus at Burleigh. One of the black swans sent by the Queen to Frogmore had the misfortune to be "shot by a nobleman's game-keeper as it was flying across the Thames." A wombat living in London made friends with the anatomist Sir Everard Home and many other of London's distinguished men of science. The great travelling menageries and fairs introduced the animals to the public at large. A kangaroo was billed at the Lyceum in the Strand as "the unparalleled animal from the Southern Hemisphere" and could be seen for one shilling. A pair of emus was included in a "most rare, varied, amusing and splendid exhibition of nature ever witnessed."

Many of the early drawings, though done with skill, still reveal the artists' difficulties in reconciling the reality before them with the doubts in their own minds. Two artists who visited Australia at the beginning of the nineteenth century stand out as painstaking draughts-men not only able to reproduce exactly what they saw but also able to fill the form with life. One was the Viennese Ferdinand Bauer, who was sponsored by Sir Joseph Banks to accompany Matthew Flinders on his voyage of circumnavigation in the *Investigator*, 1801–3. The other was Charles Alexandre Lesueur, one of the artists of the rival French expedition under the command of Nicolas Baudin. Bauer's special contribution was undoubtedly to botany, but he made a number of drawings of the animals which, for the most part, have never been published. The beautiful work of Lesueur was similarly neglected. Only a few of his paintings of the Australian animals were included in contemporary accounts of the expedition. Yet, apart from sketches, more than a hundred finished paintings on vellum, the majority of them

6a, 6b. The rare banded hare-wallaby and the extinct short-legged emu. Painted on vellum. C. A. Lesueur, of the Baudin expedition, 1800-4. *Le Havre Natural History Museum.*

of Australian subjects, are to be found in the Le Havre museum.

By the end of the century most of Australia's marsupials, as well as many other of her mammals and birds, were known. The basic similarity of structure that seemed to unite many of them caused considerable debate about how they should be classified and also stimulated discussion of the theories concerning the origin of things. Dr Smith, in the book mentioned above, analyses the influence of the Australian discoveries on the ideas that led up to the theory of evolution. In 1836 Charles Darwin himself visited Australia in H.M.S. *Beagle.* While staying near Wallerawang, he wrote in his journal: "I had been lying on a sunny bank, and was reflecting on the strange character of the animals of this country as compared with the rest of the world." He went on to consider that this might cause some people to think that "two distinct Creators must have been at work." He concluded, however, "One hand has surely worked throughout the universe."

Certainly the discovery of Australia's extraordinary animals influenced the development of scientific thought; on the other hand, the impact of European settlement on Australia's wild life was shattering, deeply disturbing the relationship of the animals to their environment. The introduction of crops, sheep, cattle and rabbits, not to mention predators such as the fox, cat, dog and wild pig, brought changes that depleted Australia's animals, and even caused the extinction of some species. The changes did not escape the notice of Charles Darwin, who, like so many visitors, was invited to a kangaroo-hunt. But the sport was bad and they did not see a single kangaroo in the course of a day's ride. "A few years since," he wrote, "this country abounded in wild animals; but now the emu is banished to a long distance, and the kangaroo is become scarce; to both the English greyhound is utterly destructive. It may be long before these animals are altogether exterminated, but their doom is fixed."

Many species, particularly those inhabiting offshore islands, where they were hunted by sealers and fishermen, were depleted rapidly. By the 1830s, for example, a variety of short-legged emu had been exterminated on both King and Kangaroo islands. Lesueur's painting (6b) and pencil sketches of such a bird form part of the scant and controversial evidence concerning their identity and location. Today, despite the introduction in the 1870s of protective laws and their subsequent revision, some species of marsupials in Australia appear to have become extinct and others are rare. Others again, such as some of the possums and kangaroos, seem to have adapted themselves to the changed environment. For some animals, such as the koala, platypus and lyrebird, the stringent enforcement of protective laws and the existence of nature reserves has meant a reprieve and a chance to consolidate numbers. Not all receive protection, however. Wherever the animals' presence conflicts with man's ever-increasing use of the land, they are controlled as pests.

Legislation, whether for protection or control, is not enough. A policy of conservation requires that we appreciate and study the complex realities that underlie a balance of nature, so that our animals will survive not merely as curiosities but as an essential part of the Australian environment.

7

The KANGAROO is Australia's largest native grazing animal. Its unique shape, upright posture and hopping movement is its own specialized answer to the Australian environment. Early observers found it difficult to reconcile these differences with the obvious similarities it bore to the grass-eating animals of other lands. To Sydney Parkinson, the artist of the *Endeavour* who first drew it (1), it appeared to be an animal "that had a head like a fawn's; lips and ears, which it throws back, like a hare's ... with a short and small neck near to which are the fore-feet, which have five toes each, and five hooked claws; the hinder legs are long, especially from the last joint, which, from the callosity below it seems as if it lies flat on the ground. ... The tail which is carried like a greyhound's was almost as long as the body, and tapered gradually to the end. The chief bulk of the animal is behind. ..." (*A Journal of a Voyage to the South Seas*, 1784.)

The kangaroo hops on its hindlegs and only goes on all fours when moving slowly, for instance while feeding. Then it rests on the forelimbs and, lifting the hindquarters off the ground, uses its tail to thrust the weight of its body forwards. The forepaws (9) are used to grasp things. Unlike the sheep, deer, and cattle it is not hoofed. The elongated hindfoot has a padded sole and a long central toe and claw. On the inside, the second and third toe have become united except for the claws which serve as a two-pronged grooming comb.

Watkin Tench observed in his *Narrative of the Expedition to Botany Bay*, 1789: "In running, this animal confines itself entirely to his hinder legs, which are possessed with an extraordinary muscular power. Their speed is very great, though not in general quite equal to that of a greyhound; but when the greyhounds are so fortunate as to seize them, they are incapable of retaining their hold, from the amazing struggles of the animal. The bound of the kangaroo, when not hard pressed, has been measured, and found to exceed twenty feet." Indeed, kangaroos have been known to maintain speeds of 25 miles an hour and when pursued can reach 30 miles an hour. The male kangaroo attains maturity at about two years. The old man of the mob can weigh more than 200 pounds and have a stand of seven feet.

The discovery of the "hopping quadruped" had caused great excitement in Europe in both popular and scholarly circles. Imagine the consternation if one of those first three specimens brought back by Banks had been a female with a pouch. After the establishment of the settlement in 1788, however, it was soon realized that kangaroos and indeed many other animals had pouches, and Thomas Pennant was able to add to his account in *The History of Quadrupeds*: "The female has on the belly an oblong pouch of vast depth. The receptacle of its young."

The doe (14), which is smaller than the male, has only one offspring at a time, though she may start breeding when she is herself only eighteen months old.

At first the settlers assumed that the animal they called Kangaroo was the same as that discovered by the first explorers well over a thousand miles to the north, on the Endeavour River. They soon realized, however, that there was a variety of animals all sharing the same general characteristics but differing from each other in size, colour and markings. In fact, all the hopping marsupials, from the smallest rat-kangaroo to the great kangaroos, are entitled to be called kangaroo, but the name is more generally used to describe the largest of these, such as the great grey forester, the red kangaroo of the plains, and the wallaroo and euro of the rocky and hilly regions.

Great grey foresters grazing (12, 13) must have been a familiar sight to the settlers, for they were one of the commonest types of kangaroo to be found near Port Jackson in those early days. It might well have been these which Captain Tench described in *A Complete Account of the Settlement at Port Jackson*: "They are sociable animals, and unite in droves, sometimes to the number of fifty or sixty together; when they are seen playful, and feeding on grass, which alone forms their food. At such time they move gently about, like all other quadrupeds, on all fours; but at the slightest noise, they spring up on their hind legs, and sit erect, listening to what it may proceed from; and if it increases, they bound off, on those legs only; the fore ones, at the same time, being carried close to the breast, like the paws of a monkey; and the tail stretched out, acts as a rudder on a ship."

The handsome red kangaroo of the plains (15–23) was not known until after the crossing of the Blue Mountains in 1813. The doe is called a "blue flyer" because of the dusty blue colour of her coat. On a winter day they can be found sunbathing in some scooped-out sand-patch.

"The elegance of the ear" wrote Tench in his *Complete Account*, "is particularly deserving of admiration:"

"this far exceeds the ear of the hare in quickness of sense; and is so flexible as to admit of being turned by the animal nearly quite round the head, doubtless for the purpose of inform-ing the creature of the approach of its enemies; as it is of a timid nature, and poorly furnished with means of defence, though when compelled to resist, it tears furiously with its fore-paws and strikes *forward* very hard with its hind legs." In this position its tail acts as a prop

supporting its weight, and as the hindlegs thrust forward, the long central claw of the foot tears down in a deadly disembowelling action. The forearms of the male are powerful enough to clasp an adversary and toss him. I have seen a fight between two large males end with one throwing the other, a good one hundred and fifty pounds of kangaroo, through the air, over his shoulder.

Young ones, at different stages of development, were found by the settlers in the pouches of the female kangaroos they shot for food. Surgeon White described some "with a young one, not larger than a walnut, sticking to a teat in this pocket. Others, with young ones not bigger than a rat. . . ." Tench observed that "it is born blind, totally bald" and that only one is born at a time, "which the dam carries in her pouch wherever she goes until the young one be enabled to provide for itself; and even then, in the moments of alarm, she will stop to receive and protect it". How it was born and came to be in the pouch eluded them. They thought, wrongly, that the young one was in some way born directly into the pouch. Pelsaert had come to a similar conclusion more than one hundred and sixty years earlier when he wrote about the strange animals he found on the Abrolhos Islands: "Below the belly the female carries a pouch, into which you may put your hand; inside this pouch are her nipples, and we have found that the young ones grow up in this pouch with the nipples in their mouths. We have seen some young ones lying there, which were only the size of a bean, though at the same time perfectly proportioned, so that it seems certain that they grow there out of the nipples of the mammae, from which they draw their food. . . ."

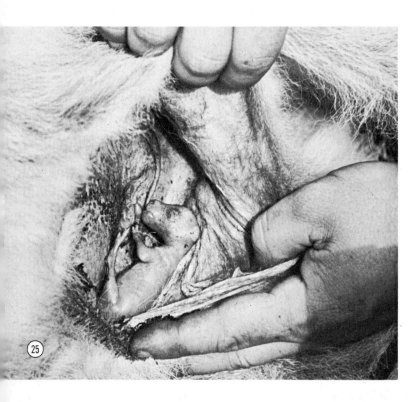

Credit for the first correct account of marsupial birth goes to an American professor who, in 1806, described how an opossum gave birth in the usual way and the young one made its unaided journey to the pouch where it attached itself to a teat. The birth of the kangaroo was first accurately described in 1830 by Alexander Collie, surgeon aboard H.M.S. *Sulphur*, in a letter to the Zoological Society, London. He described how he dissected adult females containing embryos. He also recounted the experiences of another officer who saw a doe give birth and watched as "the very diminutive young . . . crept among the fur of the mother towards her belly and towards the opening of the abdominal pouch; whilst she with her head turned towards her tender offspring, seemed to watch its progress, which was about as expeditious as a snail. After it had made some advance, my informant, unconscious of the remarkable oeconomy of generation in this class of Quadrupeds, removed the newly born

33. Probably the Kangaroo Island kangaroo. Painted on vellum. C. A. Lesueur, of the Baudin expedition, 1800-4.
Le Havre Natural History Museum.

animal before it reached its destination; which must have been the mouth of the Sac."

Considering that the newborn young is less than an inch in length it is not surprising that the birth and the progress to the pouch has so seldom been seen in its entirety. The period of gestation is variable, but averages five weeks. The mother seems to give very little assistance to the young one apart from sitting with her tail forward between her legs so that the bare underpart is uppermost. The young one has well-developed forearms which it uses to haul itself up the mother's moistened fur. Once inside the pouch it attaches itself to one of the four teats, which becomes stretched and dilated in its mouth so that it cannot be separated without damage being done. In pl. 25 the elongated teat is clearly visible and the growing joey has a certain amount of freedom of movement. It is about six weeks old. The joey in pl. 26 is about six months old. Its gawky legs stick out of the pouch as it moves about, and it can leave the teat and find it again. No nails have formed on the hands or feet, and its eyes have not opened properly.

About a month later it is ready to leave the pouch and, even then, its gradual steps towards complete independence alternate between leaning out of the pouch and nibbling the grass or having a quiet suckle with its head in the pouch and its feet on the ground. While the joey is out exercising its legs the mother will hold the pouch open and clean it; sometimes it is peremptorily tipped out for this purpose. Even after the young one has grown too big for the pouch it still receives the doe's attentive care (30, 31, 32).

Kangaroos were amongst the first Australian animals to be kept in England. There is a record, in a letter from David Dundas to Sir Joseph Banks in December 1793, of a female kangaroo in the King's collection at Richmond having given birth, ". . . the head of the young one appearing out of this Pouch as long ago as the 30th of October—since which time it has grown much, appears more frequently, and often while the mother is feeding nibbles at the Grass—sometimes one of the hind Legs and part of the Tail have been out of the Pouch at the same time with the head. The Mother is so tame she allowed me to stroke her neck with the one hand and with the other the head of the Young one." He urges Sir Joseph to come to see it, adding the postscript: "The King has not seen it yet . . . the young one has never been entirely out of the Bag—but as it has been so long in this state, it will probably soon come out. You had better therefore lose no time."

Kangaroos are very adaptable. At Whipsnade Park in England a mob of wallabies roams freely and breeds successfully (34). The only difference between wallabies and great kangaroos is one of size, particularly of the hindfoot, which in the wallabies is less than ten inches long. Together, the wallabies, the small kangaroos and the great kangaroos make up 52 of the 119 species of Australian marsupials.

The TREE-KANGAROO (36, 37) remained undiscovered in the tangled rain-forests of north-eastern Queensland until late in the nineteenth century.

Although it is not a rare animal it is not easy to find, sleeping in the treetops by day and coming down in the evenings to feed on the vegetation or perhaps visit a waterhole. In the wet season, however, it seems to find sufficient moisture in the food it eats.

The tree-kangaroo is remarkable because its typical kangaroo structure, which was originally evolved for living and grazing on the ground, has been adapted for life in the trees. Its heavy hindquarters and long hanging tail give it an awkward appearance for a tree-dwelling animal. Nonetheless it is amazingly agile. The rough pads and curved claws of its broad feet enable it to grip the branches firmly, and remedy its lack of a prehensile tail, which is a common characteristic of other tree-dwelling animals. It is able to make great downward leaps—up to forty feet from tree to ground have been recorded—stiffening its disproportionately long tail which it uses as a balancer.

The position in which the tree-kangaroo rests with its head down and tucked into its body has caused the hair on the neck and shoulders to be directed forward, thus acting as an effective watershed in the tropical downpours.

36

37

38. Ring-tailed possum. Painted on vellum. C. A. Lesueur of the Baudin expedition, 1800-4.
Le Havre Natural History Museum.

On 26th July 1770 Joseph Banks wrote in his *Endeavour* journal: "In botanising to day I had the good fortune to take an animal of the Opossum (*Didelphis*) tribe: it was a female and with it I took two young ones. It was not unlike that remarkable one which De Buffon has described by the name of Phalanger as an American animal; it was however not the same, for De Buffon is certainly wrong in asserting that this tribe is peculiar to America; and in all probability . . . the Phalanger itself is a native of the East Indies, as my animal and that agree in the extraordinary conformation of their feet in which particular they differ from all others."

This POSSUM was the only female amongst the animals that Banks caught while in New Holland; it was therefore the only one in which he discovered the presence of a pouch. Until then the only pouched animals known to naturalists were the opossum of America and the phalanger of the East Indies. Pelsaert's earlier accounts of finding animals with pouches on the west coast of New Holland had been forgotten. Yet, when the explorers returned to Europe, Bank's possum did not arouse as much interest as the more spectacular kangaroo. Once the colony of New South Wales was established, however, it was soon discovered that, like the possum, the kangaroo and many other animals had pouches.

Banks was quite right in his field observation that the animal he found at the Endeavour River more closely resembled the phalanger of the East Indies in the structure of its foot than the American opossum, although he grouped it with the latter. The nearest relatives of the American opossum amongst the Australian marsupials are not the possums of the phalanger family but the insect and flesh-eating family Dasyuridae which includes the marsupial mice, native cats and Tasmanian wolf. The word *possum* continues to be used, however, as the common name for all members of the phalanger

40. Brush-tailed possum. Woodcut. T. Bewick, *General History of Quadrupeds*, 1820.

family, including the ring-tailed possums, the brush-tailed possums, the gliders great and small, and the cuscus.

The animal captured by Banks was probably the grey ring-tailed possum of Queensland. The naturalist Thomas Pennant gave a description of it in his *History of Quadrupeds* in 1781: "New Holland Opossum, with the upper part of the head, and the back and sides covered with long soft glossy hairs, of a dark cinerous colour at the bottoms, and of a rusty brown towards the ends; belly of a dirty white. Tail taper, covered with short brown hairs, except four inches and a half of the end, which was white, and naked underneath: The skin I examined had lost part of the face; the length from the head to tail was 13 inches, the tail the same. . . ."

Species of ring-tailed possums are found over a large area of the continent and also New Guinea, ranging in colour from the grey of Queensland to the dark smoky-brown of Tasmania. The common ring-tail of south-eastern Australia (39) has a predominantly pinkish-brown back and whitish under-surface.

A Tasmanian ring-tailed possum was the only four-footed animal found at Adventure Bay when the *Resolution* called there in 1777. William Anderson, who kept a journal, remarked on the prehensile tail "by which it probably hangs on the branches of the trees as it climbs these and lives on berries." He also described well the characteristic foot structure: ". . . the feet are made much like a little hand and the hind one has a thumb distinct from the others but without a claw." This opposable first toe helps the possum to establish a firm grip. The second and third toes are united except for the claws, and this arrangement has the same grooming function for the possum as for the kangaroo.

Nocturnal in habits, the ring-tailed possum finds it hard to see in daylight, consequently its movements sometimes appear clumsy and its manner stupid. But it is well adapted to swinging around in the scrub, using its tail to anchor itself and letting its body swing forward to a new tree. Usually the female has two young ones. When they outgrow the mother's pouch they cling to her back with their tiny claws and tail and are transported in this fashion. The ring-tailed possum has to some extent adjusted itself to man.

At first it was thought that the brush-tailed possum (41, 42) found in the vicinity of Sydney Cove must be the same as that described by the early explorers, but it soon became clear that this possum with its bushy tail and fox-like face belonged to a separate species. It was on account of these features that it was first called the Vulpine Oppossum. In *The Voyage of Governor Phillip to Botany Bay* it was described from a live one already sent to England and living in the possession of John Hunter, the anatomist: "The countenance of this animal much resembles that of a fox but its manners approach more nearly those of a squirrel. When disposed to sleep, or to remain inactive, it coils itself up into a round form; but when eating or on watch for any purpose, sits up, throwing its tail behind it. In this posture it uses its forefeet to hold anything, and to feed itself. When irritated, it sits still more erect on the hind legs, or throws itself upon its back, making a loud and harsh noise. It feeds only on vegetable substances. This specimen is male. The fur is long but close and thick; of a mixed brown or greyish colour on the back, under the belly and neck of a yellowish white. Its length is about eighteen inches, exclusive of tail, which is twelve inches long and prehensile. . . ."

Thomas Bewick's woodcut of a brush-tailed possum, which he called the Squirrel Opossum (40), was based on a drawing of another one living in Durham, England.

The common grey brush-tailed possum is an active little animal at night, and in its native country it is well known in the suburbs where it can be heard scampering in the roof if one has been so unwise as to leave a gap where it can get in. Of all the marsupials it seems to have adjusted to living alongside man best, and this in spite of the fearful reduction of its numbers year after year by the skin trade. It is also the only marsupial which has a wide distribution

over the whole continent. Its favourite food amongst the indigenous plants is the peppermint gum, though it has also acquired a taste for fruit-trees and can be destructive to orchards. On the other hand, it is useful in destroying mistletoe which damages forest trees. Nothing really justifies the wholesale war made on this tame and friendly creature in the open season, mainly for the commercial value of its skin.

43

The CUSCUS (43–46) is found only at the far north-eastern extremity of Australia, on Cape York peninsular. It is related to other genera inhabiting New Guinea, the Solomons, Timor and the Celebes which were known to Europe as phalangers before the men of the *Endeavour* discovered Australia's east coast.

Aborigines brought the first cuscus to naturalist John MacGillivray, aboard H.M.S. *Rattlesnake*, while he was at Port Albany, Cape York, in 1848. In his journal he mentioned that it was "a live Opossum, quite tame and very gentle: It turned out to be a new species. . . ." But he gave no other details of its natural history.

Later tales of "monkeys" on Cape York sprang from misleading accounts of these much less energetic creatures. If disturbed during the daytime their eyes blink sleepily. The roundness of their moon-faces is unbroken by their ears which lie hidden in their woolly fur. Irregular creamy patches break up the dark areas of fur, though often these patches are so large that the light colour predominates. The long prehensile tails is hairless and scaly towards its end.

The female usually has at least one young one in its pouch and can have up to four. One I photographed was so tame that she did not mind when I placed my hand in her warm pouch.

44

45

(47)

(48)

Gliding possums were also found near Sydney in 1788. Both Governor Phillip and Surgeon-General White described the beauty of the black flying possum. Here they are represented by the two smallest, the SUGAR GLIDER (47) and the PYGMY or FEATHER-TAIL GLIDER (48). The membrane that stretches between the fore and hind limbs support them in their gliding flight from tree to tree in search of food. Unlike the large leaf-eating gliders, these two species eat a diet of insects supplemented by blossom-nectar and buds. Both occur throughout the coastal districts of the eastern and northern part of the continent.

49. Rat-kangaroo and young. Engraving from *The Voyage of Governor Phillip to Botany Bay*, 1789.

The RAT-KANGAROO was one of the first marsupials brought alive to England. *The Voyage of Governor Phillip to Botany Bay*, published in 1789, gave an illustration (49) and a description of the animal and also informed the public that two of them, one with a young one in the pouch (the models for the illustration), were to be seen at the Exeter Exchange. By all accounts they were miserable at being transferred from their natural surroundings, hiding from view in the straw of their box, and doubtless they did not survive long.

Surgeon White wrote in his journal that the native name for the animal was Poto-roo, and although it was suggested that this name should be used it continued to be called kangaroo-rat because of its structure and manner of hopping. The name Poto-roo survived, however, in the form of *Potorous*, the generic name for the long-nosed, broad-faced types. There are also several kinds of short-nosed rat-kangaroos (54).

Captain Tench detailed his observations of this creature's habits in his *Complete Account of the Settlement at Port Jackson*, 1793. "The kangaroo-rat is a small animal, never reaching, at its utmost growth, more than fourteen or fifteen pounds, and its usual size is not above seven or eight pounds. It joins to the head and bristles of a rat, the leading distinctions of a kanguroo, by running, when pursued, on its hind legs only, and the female having a pouch. Unlike the kanguroo who appears to have no fixed place of residence, this little animal con-structs for itself a nest of grass, on the ground, of a circular figure, about ten inches in diameter, with a hole on one side, for the creature to enter at; the inside being lined with a finer sort of grass, very soft and downy. But its manner of carrying the materials, with which it builds the nest is the greatest curiosity:— by entwining its tail (which, like that of the kanguroo tribe, is long, flexible, and muscular) around whatever it wants to remove, and thus dragging along the load behind it. This animal is good to eat; but whether it be more prolific at birth than the kanguroo, I know not." In fact the female rat-kangaroo usually has only one young one at a time.

Early settlers were soon familiar with this nocturnal creature. Apart from being delighted by its antics in the bush as it hopped fearlessly close to them, it was soon noticed that "it sometimes gets into the houses and lives on corn of any kind." (Anonymous note to a drawing in the Watling Collection.) This fondness for the introduced grains as a supplement to its natural foods of grasses and roots caused man to treat it as his enemy. Later, the introduction of the fox meant wholesale reduction of its numbers. Once to be found over most of Australia, except for the far tropical north, the dozen or so species are now extremely rare.

The MUSKY RAT-KANGAROO (52) has a sub-family to itself because it seems to form a link between the possums and the kangaroos, thus suggesting the line marsupial evolution has taken. All the other rat-kangaroos are like the kangaroo in not having a first toe on the hind foot, but the musky rat-kangaroo is closer to the possum in that it does have a great toe, even though it cannot oppose this to the other digits as the possum can. It also has a scaly tail similar to some members of the possum family. Together with the rest of the rat-kangaroos it has teeth which are characteristic of insectivorous animals, and indeed lives mainly on a diet of insects.

There are about thirty species of MAR-SUPIAL-MICE, varying in size from a small mouse to a large rat. The brush-tailed marsupial-rat was described by Surgeon White in 1790. The family is represented here by the FAT-TAILED MARSUPIAL-MOUSE (53), which is characterized by its very slender foot and its tail which serves as a reserve of fat. This smallest of marsupials has one of the largest litters of young; up to ten can be found in its tiny pouch at one time. In many of the other marsupial-mice the pouch is little more than a fold of skin. Marsupial-mice are most useful destroyers of pest insects such as the grasshopper, and are quite capable of extending their diet to include the common mouse.

The BANDICOOT is another small marsupial. The SOUTHERN SHORT-NOSED BANDICOOT (50) was probably the one most commonly seen round Sydney in the early days. It was first described in *The Naturalists' Miscellany*, 1797, and first given the common name of bandicoot by the explorer George Bass in 1799. It bears no relationship, however, to the bandicoot-rat of Ceylon; the use of the name was once again an attempt on the part of the Europeans to relate the strange wild life they encountered to something already known. Although its diet is a mixed one, it uses its long snout to dig out favourite grubs and other insects. It does not have a burrow, but makes a nest of grass and twigs, where it stays during the day, coming out to feed at night. Its pouch opens backwards, and it usually has a litter of four.

The BILBY or RABBIT-BANDICOOT is one of the largest of the bandicoots. It has ears like a rabbit and long silky blue-grey fur. It has powerful, well-developed forefeet which it uses to dig a proper burrow. Its chief means of protection is a strategy of escape by tunnelling one step ahead of its pursuer. It once ranged widely from east to west, but is now found only in the arid inland regions and the south-western corner of Australia.

Many of these small animals represent some of the most interesting marsupial adaptations; despite this they are often thoughtlessly destroyed, and they are liable to become extinct unless there is more widespread understanding and protection.

53

54

55. Wombats. Painted on vellum. C. A. Lesueur, of the
Baudin expedition, 1800-4.   *Le Havre Natural History Museum.*

"Dear Sir,

Having lately met with an animal found upon that Island on which the Ship *Sydney Cove* was wrecked, and which appears to me to be quite a new Animal, I have preserved it in spirits to send it to the Literary and Philosophical Society lately established at Newcastle upon Tyne. . . . This is the only opportunity I have had of picking up anything worth the notice of the gentlemen of that Society; I have taken the liberty of addressing it to you, that you might have an opportunity of examining it; I have made a little drawing of it whilst it was alive, which I send with it. . . . the Wombat, which is the name the natives of the mountains give to the above animal, they have seen in the interior of this country, and are said to be delicate meat. You will also receive Sir, a Box, in which is contained a Skin of the Wombach, and the Bones of the Head. . . ." (A letter from Governor Hunter to Sir Joseph Banks, 5th August 1798.)

Very few new animals had been found in the years which followed the first startling wild life discoveries, for they were years of struggle against starvation, disease, fear and isolation. As the little colony gradually consolidated, the Europeans reached out to extend its boundaries both inland and along the coast. The WOMBAT was one of several new animals found about that time. Its tender flesh saved the lives of the men shipwrecked on Preservation Island in July 1797, and again in January 1798, a wombat was found by the searchers for a rumoured "land of plenty" to the south of Sydney. Governor Hunter's sketch of the wombat inspired the first published illustration of the animal in Judge-Advocate Collins's *Account of the English Colony in New South Wales* in 1802. Another charming drawing (55) was made by Charles Alexandre Lesueur, the French artist of the Baudin expedition, 1800–4. His otherwise fine picture shows four offspring instead of the usual one.

Most observers thought the wombat resembled either the beaver, badger or bear but fortunately it continued to be called by its native name. "The Wombat is a squat, thick, short-legged and rather inactive quadruped," wrote David Collins, "with a great appearance of stumpy strength, and somewhat bigger than a large turnspit dog." Like the koala it is tailless. It has the strong forefeet of a burrowing animal. A bark nest is made at the end of the burrow where it rests during the day, although it is not entirely nocturnal. The biggest variation between the different species of wombat is one of size. The dark brown Tasmanian wombat weighs about 40 pounds and is only half the size of those found on the mainland, in the hilly or mountainous coastal areas.

David Collins recounted the explorer George Bass's experiences with the wombat: "This animal has not any claim to swiftness of foot, as most men could run it down. . . . In disposition it is mild and gentle, as becomes a grass-eater; but it bites hard, and is furious when provoked. Mr Bass never heard its voice but at that time: it was a low cry between a hissing and whizzing, which could not be heard at a distance of more than 30 to 40 yards. He chased one and with his hand under his belly suddenly lifted him off the ground without hurting

57

58

him and laid him upon his back along his arm, like a child. It made no noise, nor any effort to escape, not even a struggle. Its countenance was placid and undisturbed, and it seemed as contented as if it had been nursed by Mr Bass from its infancy. He carried the beast upwards of a mile, and often shifted him from arm to arm, sometimes laying him upon his shoulders, all of which he took in good part; until, being obliged to secure his legs while he went into the brush to cut a specimen of new wood, the creature's anger arose with the pinching of the twine, he whizzed with all his might, kicked and scratched most furiously and snapped off a piece from the elbow of Mr Bass's jacket with his grass-cutting teeth. Their friendship was there at an end, and the creature remained implacable all the way to the boat. . . ."

The tame and playful wombat is not easily found these days. It proved too defenceless. Visiting sealers to the islands of Bass Strait soon reduced the numbers of the Tasmanian species, and like those of the mainland they survive only in isolated colonies.

63. Koala with young. Water-colour. F. Bauer,
artist aboard the *Investigator*, 1801-3.

Sydney Gazette, 21st August 1803:
"An animal whose species was never before found in the Colony, is in His Excellency's possession. When taken it had two Pups, one of which died a few days since. This creature is somewhat larger than the Waumbut, and although it might at first appearance be thought much to resemble it, nevertheless differs from that animal . . . the graveness of the visage . . . would seem to indicate a more than ordinary portion of animal sagacity. . . . The surviving Pup generally clings to the back of the mother, or is caressed with a serenity which appears peculiarly characteristic; it has a false belly like the appossum and its food consists solely of gum leaves, in the choice of which it is excessively nice."

The KOALA so described was the one which the Ensign Francis Barrallier, of the N.S.W. Corps, had brought alive to Governor King. It must have been about this time that the artist Ferdinand Bauer, who had accompanied Flinders on his voyage of circumnavigation, made some drawings of this relatively new find (63). They show the same exquisite attention to detail and feeling for the animal's nature as do the other drawings he made of Australian animals. None of them, however, were published in contemporary accounts.

The koala had been seen for the first time several years earlier, on that same journey south of Sydney in January 1798 when the wombat and lyrebird had been found. One of the party, John Price, wrote: "There is another animal which the natives call a Cullawine which much resembles the sloths of America." It was also called *Colo, Koolah* and *Koala,* all aboriginal names which are supposed to refer to its ability to live without water. The koala chooses its exclusive diet of gum-leaves from only twelve species of eucalypts. It eats about $2\frac{1}{2}$ pounds a day, and to help digest such a large quantity it has an extra-long appendix of six to eight feet.

Somewhere in the course of marsupial evolution the koala has got out on a limb by itself. Like the wombat, the koala has a stumpy, bear-like shape and no tail, but it has taken to living in trees where its lack of a prehensile tail is made up for by its strong limbs and sharp claws. It is most closely related to the tree-dwelling possums, and like them has a hand with two digits opposed to the rest and a hindfoot with an opposable, nailless great toe, so useful for gripping branches firmly. The female usually has only one offspring at a time, which she carries in her pouch for six to eight months; after that the young one clings to her back.

Clearly an animal with such specialized habits could not survive long once its environment had been disturbed by the coming of the European. In the early days it must often have ended up in the cooking pot, and as settlement advanced and the forests were destroyed its feeding grounds were reduced. Worse still, slaughter of this animal for its skin began, which, combined with an epidemic, brought it to the verge of extinction. A short open season in 1927 in Queensland, resulted in the export of half a million skins. But for the Fauna Protection Act the little zipped-up toy koala bear which represents Australia all over the world would soon have been the only reminder of the one-time existence of this unique animal. Now its survival is at least secure in the nature reserves of the eastern coast from Victoria to Queensland, and in the forests of Victoria the koala has begun to make up its numbers again.

65

67. Native cats with a fowl. Water-colour. Anonymous artist, Watling Collection, 1790s.

The NATIVE CAT (68, 69) has the distinction of being one of the three animals which the men of the *Endeavour* collected as specimens in 1770. The others were the possum and the kangaroo. Joseph Banks wrote of it: "Another was calld by the natives *Je-Quoll*: it is about the size and something like a polecat of light brown spotted with white on the back and white under the belly." Apart from the kangaroo, it was the only animal from New Holland which the artist of the *Endeavour*, Sydney Parkinson, drew. It is a faint pencil sketch, and with it is a description of the animal as a "spotted one of the viverra kind", referring to its carnivorous, weasel-like ways.

Spotted Opossum was one of the names given it by the settlers of 1788, who noticed that it varied in colour from light brown to black, with creamy spots on the body and no spots on the tail. Now known as the common native cat, it is about the size of a domestic cat. They also saw another animal rather like it, except that it is about twice its size and has spots on the tail as well as the body. This was known as the Spotten Marten and is now called the tiger cat. Yet the name *cat* is just as misleading as *marten* and other names derived from observations of the behaviour of the animal rather than its structure. For they are both pouched animals and share a common marsupial ancestor with the rest of Australia's marsupials. From the first small insectivorous creatures which arrived on the continent the whole diversified range has come. The native cats and tiger cats, together with the marsupial mice at one end of the size-scale and the Tasmanian wolf at the other, make up the family Dasyuridae. They are probably closest in structure to their primitive ancestors both in the possession of front incisor teeth and in the separated toes of the hindfoot. The members of this family are the only marsupials which can be called predatory, each one preying on living creatures within its size range from insects to mice and rats, reptiles, birds and small mammals. A tiger cat has been known to attack a small wallaby.

The ferocious flesh-eating habits of the animal were discovered early, and in the Watling Collection there is a drawing of two native cats, one dark and the other light, with a dead fowl (67). Habits like these have resulted in its widespread destruction by man, although many of the smaller creatures it feeds on are pests. Different varieties of native cat were once to be found over most parts of Australia, except Tasmania. Their territories are now greatly restricted, although they may still be found near Sydney.

The pouch of the native cat is little more than a fold of skin and the opening faces backwards. As many as twenty-four offspring may be born at one time but, as there are only six teats for them to attach themselves to, only six can survive. The native cat shelters in caves or in hollow logs and is generally a nocturnal, tree-climbing, though not tree-dwelling, animal. Its larger relative, the Tasmanian wolf or tiger, which parallels the non-marsupial dog of other parts of the world, is now almost extinct, and found only in the fastness of the Tasmanian forests. It was probably driven from the Australian mainland by the non-marsupial dog, the dingo, which came to the continent with the aborigines.

68

69

70. "A Large Dog". G. Stubbs. This painting was exhibited next to the painting of the kangaroo in 1773.

Mrs W. P. Keith, London.

The presence of the DINGO or wild dog was guessed at by William Dampier on his first visit to the west coast of New Holland in 1688 when he found the tracks of "a Beast as big as a Mastiff-Dog". On his second visit in 1699 he wrote that "my men saw two or three Beasts like hungry Wolves, lean like so many skeletons, being nothing but Skin and Bones: 'Tis probable it was the Foot of one of those Beasts that I mention'd as seen by us in N. Holland. . . ."

Almost a hundred years later, on the eastern coast, both Cook and Banks mentioned several sightings of an animal like a dog or wolf. The most emphatic of these was made by Banks on 29th June 1770 at Endeavour River: "One of our Midshipmen an American who was out a shooting today saw a Wolf, perfectly he sayd like those he had seen in America; he shot at it but did not kill it." It seems that Banks did not succeed in securing a specimen because he does not include the wild dog in his list of animals whose skins he collected. What then did the artist George Stubbs use as a model for his painting of "A Large Dog" (70) which he did for Banks and which was shown next to "A Portrait of the Kongouro from New Holland, 1770", in an exhibition of the Society of Artists, London, in 1773? The erect ears, bushy tail and general features are so clearly those of a wild dog that it seems reasonable to accept that it was the wild dog of New Holland which the artist had in mind; for in those years there was no more exciting topic of conversation in society and amongst naturalists than the findings of Mr Banks and Dr Solander. The wild dog seen by the explorers was certainly discussed with the naturalist Thomas Pennant, who in 1781, in *The History of Quadrupeds* followed a description of the wild dog—class Lupus—as found in Asia, Europe and America, with the statement "and are believed to inhabit *New Holland*; animals resembling them have been seen there by the late circumnavigators . . . Dampier's people also saw some. . . ."

By 1793 Pennant was able not only to enlarge on this information in a new edition of his book, drawing on both the account based on Governor Phillip's papers and Surgeon White's journal, but to refer to two dingoes living in England since 1789. He described the animal as having "short erect sharp-pointed ears: a foxlike head; colour of the upper part of the body pale brown; grows lighter towards the belly; hind part of the forelegs, and fore part of the hind legs, white: feet of both of the same color; tail very bushy: length about two feet and a half: of the tail not a third of the body: height about two feet. Inhabits New Holland and seems the unreclaimed dog of the country. Two have been brought alive to *England*; are excessively fierce, and do not show any marks of being brought to a state of domesticity. It laps like other dogs; but neither barks nor growls, when provoked; but erects its hairs like bristles, and seems quite furious. It is eager after its prey and is fond of rabbits and fowls, but will not touch dressed meat: is very agile: It once seized a fine *French Dog* by the loins, and would have soon destroyed it had not help been at hand. It leaped with great ease on the back of an ass and would have worried it to death, had not the ass been relieved, for it could not disengage itself from the assailant. It was known to run down deer and sheep."

(71)

The dingo has sometimes interbred with the domestic dog gone wild. It displays a considerable cunning and skill in hunting its quarry, and can be very destructive, attacking large numbers of sheep apparently for the sheer sake of killing. Man has made extraordinary efforts to defeat and exterminate the dingo; thousands of miles of special dingo-proof fencing have been erected in great lines across the continent to try to confine it to non-pastoral areas. High bonuses are paid, both by the government and the pastoralists, for every dingo scalp brought in, and there are people whose sole occupation is dingo-hunting, or "dogging".

(72)

73. Platypus. Water-colour. F. Bauer, artist aboard the *Investigator*, 1801-3.

A PLATYPUS was Sir Joseph Banks's choice of a gift for Napoleon in 1802, and a record of how the great man received the news has survived in a letter from the physician Sir Charles Blagden to Sir Joseph:

"La Place then very judiciously said, 'Sir Joseph Banks is sending a most curious animal I have mentioned to you; partaking of the structures of different classes.' 'Yes,' said the First Consul, 'he has always thought of us; what are the classes of which it partakes?' I answered, 'The quadruped, bird and lizard.' He asked its name: and as he boggled at it a little, I explained its meaning in Greek: Ornithorhynchus Paradoxus."

Well might Napoleon boggle. Of all the wild life from the new colony which astounded Europe, nothing must have seemed more unlikely than the platypus. Even the scientists were inclined to disbelief and maintained that, like "the eastern mermaid", this furred animal with its duck-like bill and webbed feet, must be a fake product of the "artful Chinese".

The first platypus to reach England had been sent by Governor Hunter to the Literary and Philosophical Society of Newcastle upon Tyne, by way of Sir Joseph Banks. The Governor, in writing about the wombat to Banks, mentioned including "a drawing of an amphibious animal lately found in the fresh water lakes; the skin of the little creature which is of the mole kind, is in the bag containing the Wombat. . . ." (Sydney, 5th August 1798.)

Governor Hunter's drawing served as the model for both the first published engraving of the animal in David Collins's journal in 1802 and Thomas Bewick's woodcut of "an amphibious animal" which reached a wider public in his 1820 edition of *The General History of Quadrupeds*. Another early unpublished painting (73) by Ferdinand Bauer was probably done between 1802 and 1805 while Bauer was in Australia. In this painting, as in many other early representations, there appears to be a frilly edge to the bill where it meets the body. This would suggest that Bauer used as his model a stuffed animal in which the flap of cuticle at the base of the bill had dried and contracted. In the living animal this flap overlaps the hair about three-quarters of an inch and lies loosely and smoothly upon it.

The platypus's habitat ranges over the eastern half of Australia from the Atherton Table-

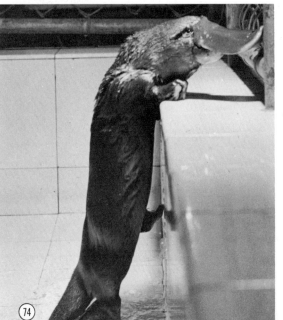

land to Tasmania. It frequents mountain streams, lagoons and waterholes. The first platypus was found on the banks of a lake near the Hawkesbury River and was described by David Collins in his journal: "In size it was considerably larger than the land mole. The eyes were very small. The fore legs, which were shorter than the hind, were observed, at the feet, to be provided with four claws, and a membrane, or web, that spread considerably beyond them, while the feet of the hind legs were furnished, not only with this membrane or web, but with the four long and sharp claws, that projected as much beyond the web, as the web projected beyond the claws of the fore feet.

The tail of this animal was thick, short and

very fat; but the most extraordinary circumstance observed in its structure was, its having, instead of the mouth of an animal, the upper and lower mandibles of a duck. By these it was enabled to supply itself with food, like that bird, in the muddy places, or on the banks of the lakes, in which its webbed feet enabled it to swim; while on shore its long and sharp claws were employed in burrowing; nature thus provided for it in its double or amphibious character. These little animals had been frequently noticed rising to the surface of the water and blowing like a turtle."

A long while elapsed before scientists agreed that this small amphibious creature was a true mammal. The anatomist Sir Everard Home, dissected a male and a female and read a careful and accurate description of his findings to the Royal Society in 1801, but he did not detect the female's milk glands. These were discovered much later in 1824 by the German anatomist Meckel, and it was not until the 1880s that the platypus's egg-laying nature was indisputably established. It is this last characteristic which distinguishes the platypus and the spiny ant-eater from other mammals. They are both furred, warm-blooded and suckle their young like other mammals, but they lay eggs in the manner of birds or some reptiles. Because of this it was suggested that these two animals be placed together in a distinct class between birds and mammals, called monotremata—a name which refers to the organs of generation and elimination sharing a common outlet. Nowadays they are still known as Monotremes and are the only living survivors of this most primitive form, but they are included within the broader classification of mammals.

The first platypus to breed in captivity was at Healesville Sanctuary in Victoria in 1943, under the care of David Fleay. Since 1956 the new Platypusary at Healesville, with its carefully designed artificial burrows and large glass-sided tank, has introduced this shy creature to thousands of people as well as facilitated closer scientific study of its habits. Here one can watch the aquatic ballet as it dives and nuzzles the bottom with its tough but pliable snout in its search for food. Under water the platypus relies on this sensitive bill and keeps its eyes and ears closed. It rises to the surface every minute or so to breath. In captivity it will take food from the hand (78).

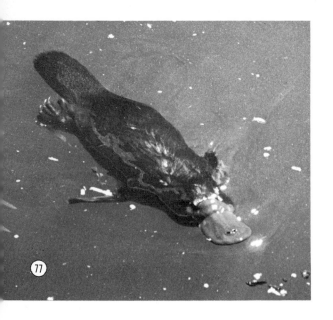

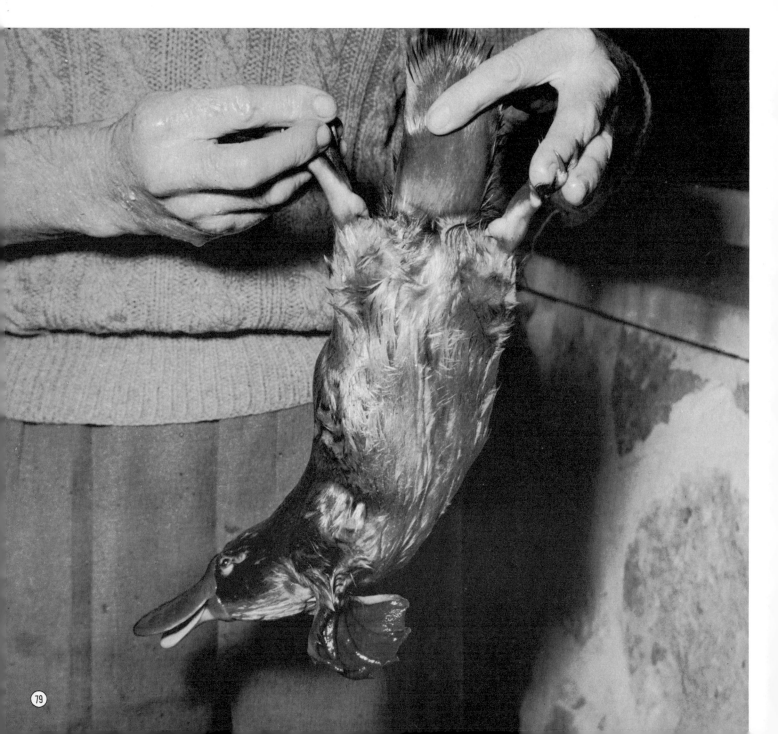

The platypus makes its burrow in the bank of a stream. It spends several hours a day in the water, while feeding in the early morning and again in the evening. When out of the water the webbing of the forepaws folds back and leaves the claws free for walking or digging. A special brooding burrow is excavated by the female to which she retires to lay her eggs and there she remains until they are hatched. She plugs the entrance to the burrow with a succession of earth barriers which have a protective function and also serve to keep the temperature constant. She usually lays two (sometimes three) small eggs —about half-inch, soft-shelled and ivory-coloured—which take about ten days to hatch out. She feeds her offspring with milk that oozes from enlarged pores. The young ones remain in the nest for about four months.

Why the male is equipped with a poison spur on each hind limb (79) is uncertain. With these he is able to inject venom and has been known to do so when mishandled by man. He is an inoffensive animal, however, with few natural enemies except perhaps the goanna and large snakes.

Once hunted for its skin, the platypus is now rigorously protected so that even possession of a skin can result in a heavy fine. One of the present-day threats to its survival is probably the practice of fishing with traps. When the platypus is caught in one of these it cannot rise to the surface to breathe, so it drowns. Rabbit warrens that riddle the banks of many streams possibly discourage the platypus from digging its burrow, particularly the brooding burrow.

While it is amusing and instructive to watch these little animals in the platypusary it is even more rewarding to chance upon them, as I once did, playing fearlessly and freely in a pool at the foot of a waterfall.

(83)

(84)

(85)

The quills of the SPINY ANT-EATER suggest a resemblance to the rodent porcupine, so that at first it was mistaken for this animal. Its unique place in the scheme of things, as an egg-laying mammal like the platypus, was not realized.

The natives in the vicinity of Sydney called it *Burro gin* and its habits and structures were well described in a note (probably written by Surgeon White between 1792 and 1795) to an early unpublished drawing in the Watling Collection: "The natives informed me this Animal . . . tho' rarely seen by us is preety numerous in the interior parts of the Country; they add that the flesh they consider is a great delicacy; that the Animal lives a good deal on Ants, being mostly found in the neighbourhood of their Hills . . . but that they principally live on Dew which they lick in with a red flesh Tongue well fitted to their Extraordinary Small Bill Mouth. In the daytime we seldom or ever have seen any of them; the Natives say in the Nights, or very late Evening, or early in the morning they may be discovered (at least their Haunts by a constant single Whistle which the Natives will imitate, and by that means surprise them, before they discover their danger, or can get off (for their gait and movements are slow and heavy) or burrow themselves in the Earth. . . ."

This defensive burrowing action (83–86) was also mentioned by David Collins, who recounted how Mr Bass's dogs found a "porcupine ant-eater" at Twofold Bay, "but they could make no impression on him; he escaped from them by burrowing in the loose sand, not head formost, but sinking himself directly downwards, and presenting his prickly back opposed to his adversaries".

The spiny ant-eater is found over most of the Australian mainland, New Guinea and Tasmania. Differences are exhibited in size, colour, length of snout, relationship between length of spines to hair, and length of claws. The long-quilled varieties have an extra-long curved claw on the second toe to groom between the spines. Unlike the platypus which lays its eggs in a burrow, the spiny ant-eater has a pouch into which the egg is manoeuvred by bodily contortions. Here it is hatched and the young is suckled and carried until its tiny spines make such an arrangement inconvenient.

86

87

88

89. Black swan. Water-colour. The Port Jackson Painter, Watling Collection, 1790s.   *British Museum (Natural History)*.

"We found two young swans on the river which we overtook by rowing fast, caught them with a hook which slightly injured one of them whereas the other was not hurt, and brought them aboard. They are quite black." Thus wrote the explorer Willem de Vlamingh on 7th January 1697. He named the river *Swaate Swaane Drift* after the birds he found there, and took three of them with him to Batavia, the first Australian wildlife to be exported. The BLACK SWANS (90) did not survive to reach Europe, but news of them did. In 1698 the English Dr Martin Lister received a letter from Holland describing their discovery, and the contents were published in the *Philosophical Transactions of the Royal Society*. The official artist of the expedition, Victorzoon, made some paintings of the west coast of New Holland, and although these seem to have been lost, one which showed black swans and a rowboat on the river was reproduced in a book called *Oudt en Nieu Oost-Indien* by Rev. François Valentjin published in 1724-6.

On the east coast the settlers of 1788 found black swans to be plentiful. One of the first exploring parties saw them on the Deewhy Lagoon, just north of Manly. The occasion was described in the account of the colony based on Captain Phillip's papers: "On this lake they first observed a black swan, which species, though proverbially rare in other parts of the world, is here by no means uncommon, being found on most of the lakes. This was a very noble bird, larger than the common swan, and equally beautiful in form. On being shot at, it rose and discovered that its wings were edged with white: the bill was tinged with red." One of the finest paintings in the Watling Collection is of a black swan (89) done in the early 1790s by an anonymous artist referred to as the Port Jackson Painter.

The black swan is a gregarious bird, and David Collins, in his journal, mentions that while he was at Port Dalrymple the explorer George Bass found an estimated "300 swimming within the space of a quarter of a mile square". It is still to be found in large numbers in the south and east where it manages to co-exist with man; and in the west, where it was first seen, it has become the emblem of the State of Western Australia.

It is not surprising that water-birds were the wildlife most frequently described before settlement, for the inhospitable nature of the country and the apparent lack of any big rivers limited the contacts of the early explorers to the coastline on both the east and the west side of the continent. The AUSTRALIAN PELICAN (91) was noticed by them all. While at Botany Bay, Lieutenant Cook wrote in his journal: "Water fowl are in no less plenty about the head of the harbour where there are large flats of sand and Mud on which they seek their food, the most of these were unknown to us, one sort specially which was black and white and as large as a goose but most like a pelican."

Water-birds, particularly oceanic wanderers and waders, are the birds which Australia shares most freely with other parts of the world. There are many sorts of birds, however, particularly land-birds, which are not represented in Australia at all, and a large percentage of the birds which are found here are found nowhere else.

The BRUSH-TURKEY (92, 93) and the other mound-builders, the scrub-fowl and the mallee-fowl, are found only in Australia. The mound, which can be some 20 feet round and 6 feet high, is built by the male. The eggs are buried about 2 feet down and the heat generated by the decaying twigs and leaves incubates them. From time to time the male bird digs into the mound-top and puts his head down to check the temperature.

94. Emu. *The Voyage of Governor Phillip to Botany Bay*, 1789. An engraving derived from the first on-the-spot drawing by Lieutenant Watts.

The EMU was discovered within a month of the arrival of Captain Phillip and his party at Port Jackson. The various chroniclers of those early years, Captain Phillip, Surgeon White and Captain Tench, all mention the occasion. The bird was shot by a convict within two miles of the settlement. "Its weight, when complete was seventy pounds, and its length from the end of the toe to the tip of the beak, seven feet two inches, though there was reason to believe it had not attained its full growth," wrote Tench in his *Narrative of the Expedition to Botany Bay*. A drawing was made on the spot by Lieutenant Watts. Though crudely done it was a lively representation and, reproduced as an engraving (94) in the account of the colony based on Captain Phillips' papers, it served to introduce the New Holland cassowary, as it was first called, to Europe. Later it became known as the Emu, a name derived from the Portuguese *Ema*, which simply means a large bird.

The skin of this first specimen was preserved, "put into attitude" and sent to Lord Sydney, who gave it to Sir Joseph Banks who, in turn, passed it over "to be deposited in the collections of natural history of Mr John Hunter, Leicester Square". This was the famous anatomist who made the first detailed descriptions of a number of the new animals sent from the colony. Here it was seen, examined and drawn by the interested naturalists, including the ornithologist John Latham, who published the first descriptions of so many Australian birds.

In his *Journal of a Voyage to New South Wales*, Surgeon White declared that the bird was "in every respect much larger than the common cassowary of all authors, and differs so much therefrom, in its form, as to clearly prove it a new species." He then proceeded to outline the similarities and differences. "The colour of the plumage is greatly similar, consisting of a mixture of dirty brown and grey; on the belly it was somewhat whiter; and the remarkable structure of the feathers, in having two quills with their webs arising out of one shaft, is seen in this as well as the common sort. It differs materially in wanting the horny appendage on the top of the head. The head and beak are much more like those of the ostrich than the common Cassowary, both in shape and size. Upon the upper part of the head the feathers, with which it is but thinly covered, are very small, looking more like hair than feathers, and in having the neck pretty well clothed with them, except the chin and throat, which are so thinly covered that the skin, which is there of a purplish colour, may be seen clearly. The small wings are exceedingly short, which form a ridiculous contrast with the body, as they are even less than those of the Cassowary: they have no large quills in them, being only covered with the small feathers that grow all over the body."

The unusual structure of the feather in which two long slender shafts issue from the one quill, was remarked on by all the early observers, and is characteristic of the two large flightless species, the emu and the cassowary. In his second book, *A Complete Account of the Settlement at Port Jackson*, Tench noted that the natives denied the emu to be a bird, "because it does not fly". The emu's protection is its excellent camouflaged coat, its long legs and great speed and its not insignificant weight, power and ability to kick and peck if cornered. It was described as shy but it is also exceedingly curious, and it is surprising that there are no early stories about the antics it will indulge in to investigate anything which excites its interest such as a flapping piece of cloth.

Of the bird's legs White wrote: "As to the back part of them, the whole length is indented, or sawed, in a remarkable manner. The toes are three in number, the middle one long, the other two short, with strong claws. . . ." This structure of the leg led Tench to think that, although the emu was quite common around the settlement it must have come from a rather different environment originally. "The legs which were of vast length", he wrote, "were covered with strong scales, plainly indicating the animal to be formed for living amongst deserts . . ." and it is in the semi-arid regions of Australia's outback that the emu can still be found in considerable numbers.

As well as the common emu of mainland Australia there was once an emu of shorter stature and darker plumage which lived on King Island and probably also on Kangaroo Island. Whether there was more than one species of short-legged emu remains uncertain, for by the time South Australia was settled in 1836 the birds were no longer to be found on either island; sealers and fishermen had wiped them out. Most of the evidence for their existence derives from accounts, drawings and specimens brought back by the French expedition, led by Nicolas Baudin, which explored Australia's southern coastline from 1800 to 1804. In the Paris Museum of Natural History there is a stuffed short-legged emu labelled as from King Island. This was one of two which survived the return voyage to France. There they were kept in the menagerie until they died in 1822.

The official account of the expedition by Péron and Freycinet included an engraving of two short-legged emus said to be from Kangaroo Island. The original drawing (6b) upon which this engraving was based was painted on vellum by C. A. Lesueur, one of the artists of the expedition.

Captain Tench also gave an account of the finding of emu chicks (105, 106) for the first time. He wrote: "The largest cassowary ever killed in the settlement weighed ninety-four pounds: three young ones which had been by accident separated from the dam, were at once taken, and presented to the governor. They were not larger than so many pullets, although at first sight they appeared to be so, from the length of their necks and legs. They were very beautifully striped, and from their tender state, were judged to be not more than three or four days old. They lived only a few days."

Tench went on to describe how, on other occasions, both the emu eggs and a nest (101-104) were found: "A single egg, the production of a cassowary, was picked up in a desert place, dropped on the sand, without covering or protection of any kind. Its form was nearly a perfect ellipsis; and the colour of the shell a dark green, full of little indents on its surface. It measured eleven inches and a half in circumference, five inches and a quarter in height, and weighed a pound and a quarter.—Afterwards we had the good fortune to take a nest: it was found by a soldier, in a sequestered solitary situation, made in a patch of lofty fern, about three feet in diameter, rather of an oblong shape, and composed of dry leaves and tops of fern stalks, very inartificially put together. The hollow in which lay the eggs, twelve in number, seemed made solely by the pressure of the bird. The eggs were regularly placed. . . . The soldier, instead of greedily plundering his prize, communicated the discovery to an officer, who immediately set out for the spot. When they arrived there, they continued for a long time to search in vain for their object; the soldier was just about to be stigmatized with ignorance, credulity, or imposture, when suddenly up started the old bird, and the treasure was found at their feet." It was fortunate that, on this occasion, curiosity took precedence over hunger. Undoubtedly many new animals must have gone unrecorded in the first days of settlement because they were eaten.

After laying the eggs the female has nothing more to do. They are incubated by the male bird who fiercely protects the nest. After some sixty days the chicks hatch out. Their stripes, an excellent protection, are retained for about ten weeks.

The emu was often killed for meat and was favoured because its flesh "tasted not unlike young tender beef". It was while on a hunting expedition that Tench saw adolescent emus (107–109): "I have nevertheless had the good fortune to see what was never seen but once, in the country I am describing, by Europeans —a hatch, or flock, of young cassowaries, with the old bird. I counted ten, but others said there were twelve. We came suddenly upon them, and they ran up a hill, exactly like a flock of turkies, but so fast that we could not get a shot at them." It is two years before the young emus reach maturity.

The emu was once found over the whole Australian mainland. Inroads on its numbers by hunting and disturbance of its haunts caused it to retreat as settlement advanced. As early as 1822 Lesson, the French medical officer aboard *La Coquille*, had noted that "Already the emeu no longer inhabits the plain called by its name and which it formerly filled . . . [it] has fled beyond the Blue Mountains or beyond the limits of the Cow Pasture." On the plains, in the same habitat as the red kangaroo, the emu can still be found, sometimes in small groups, sometimes in large flocks. It feeds on native fruits, vegetation and insects as well as cultivated crops. The wheat farmer regards it as a pest because of both the grain it eats and the damage it causes by trampling the crops.

This is particularly the case in the wheat belt of Western Australia, north-east of Perth, where, in 1932, a most extraordinary campaign to destroy the emus was undertaken; it became known as "the emu war". The army was called in to machine-gun the birds. The poor score was an embarrassment to the army, and a public outcry was raised at the choice of such shameful methods. In fact, the rabbit fences built across the State did much more to limit the emus and, together with more conventional means of control, accounted for the death of many thousands of birds. In 1935, when the bonus paid on an emu's head was a shilling, 57,034 birds were destroyed between August and the following January. The slaughter has continued, and 39,367 birds were killed in 1960. The heaps of emu bones strewn along the No. 3 rabbit fence which runs east-west through the township of Ajana, near Geraldton, is a silent comment on the efficacy of this fence in arresting the seasonal movement of the emus southward.

111

112

The CASSOWARY (111-113) remained undiscovered in the tropical forests of north-eastern Australia until the tragic expedition led by the explorer Edmund B. C. Kennedy attempted to penetrate the area in 1848. Of the thirteen men who left H.M.S. *Rattlesnake* at Rockingham Bay only two were to survive. One of these was the botanist William Carron who later wrote an account of the journey. He told how a cassowary was shot by Jackey, the aboriginal with the party. His description of the bird was included in a letter from John MacGillivray of the *Rattlesnake* to John Gould, the famous naturalist, in 1849. He wrote: "The cassowary is gaudily covered with blue about the head, and furnished with a helmet. Carron the botanist, one of the survivors of the Kennedy expedition, told me this, adding that Wall thought so much of his prize that he carried the skin on his back until they arrived at Weymouth Bay where he died." The first published account of the new find was written by Thomas Wall's brother, William Sheridan Wall, curator of the Australian Museum, for the *Illustrated Sydney News*, in 1854. No actual specimen was brought back until 1866 when another cassowary was shot near Rockingham Bay.

Like the emu, the cassowary is flightless. Though shorter than the emu, it weighs much more, the largest emu being about 150 pounds and the average cassowary ranging from 160 to 200 pounds. The feathers are a deep, rich, glossy black, sparse on the neck where the skin shows through a bright blue. The cassowary has red wattles and a tough, horned helmet. In 1874 the Sydney ornithologist E. P. Ramsay, who wrote a life history of the bird, recorded that though "tolerably plentiful only a few years ago" in the area between Rockingham Bay and the Endeavour River, the arrival of the sugar planters had resulted in "these fine birds [being] most ruthlessly shot down and destroyed for their skins, several of which I saw used for hearth rugs and door mats. . . . I know of no bird so wary and timid; and although their fresh tracks may be plentiful enough, and easily found in the soft mud on the sides of the creeks, or under their favourite feeding trees, yet the birds themselves are seldom seen . . . they not infrequently 'show fight' by bristling up their feathers and kicking out sideways or in front with a force sufficient to knock a man down . . . the sharp nail of the inner toe is a most dangerous weapon, equal to the claw of a large kangaroo, and capable of doing quite as much execution. In traversing the scrub the head is carried low to the ground, and the vines and branches of the trees, striking the helmet, skid over it onto the back."

114. The first illustration of a kookaburra, described as *Le Grand Martin Pêcheur de la Nouvelle Guinée*. P. Sonnerat, 1776.

The KOOKABURRA (115, 116) is perhaps the best known of all Australian birds. It was first described and illustrated (114) in 1776, by Pierre Sonnerat in his *Voyage à la Nouvelle Guinée*, as *Le Grand Martin Pêcheur*, the largest species of kingfisher yet known. For many years it remained a mystery as to how Sonnerat came to include this specifically Australian kingfisher in his list of birds from New Guinea, for he at no time touched on the Australian coast and paid only a cursory visit to New Guinea. One theory was that Malay trepang-fishers must have taken the birds to New Guinea from the Australian mainland. In 1956, however, the naturalist Dr Averil Lysaght discovered a long-overlooked letter to Joseph Banks from Pierre Sonnerat in which the latter recalled: "I had the honour of seeing you at Cape of Good Hope in 1770, you were kindness itself, and you gave me some new birds which you had found in that country which you had just travelled through. . . ." So this was probably how Sonnerat came by a specimen of the kookaburra, which he wrongly attributed to New Guinea. The elongated shape of the bird in the engraving certainly suggests that the latter was made from a skin and not from a sight of the live bird.

John Latham described a Great Brown Kingfisher in his *General Synopsis of Birds*, and although he acknowledged seeing two specimens of the bird in Bank's collection as well as Sonnerat's illustration, it did not occur to him to question Sonnerat's claim, so he stated that the bird was from New Guinea "whence Sonnerat had the bird figured by him"; so the error was perpetuated.

The first settlers quite rightly thought that the large brown bird which they saw around Sydney Cove was the same great brown kingfisher. A drawing of it in the Watling Collection is accompanied by an anonymous note: "This Bird lives on Insects, Worms etc. . . . from its Note which is that of a human loud and continued Laugh, it might be considered a chearfull Bird. . . . It is a Bird of slow and short flight; and seems when on the Wing to have some difficulty to support its fore-part, which regularly from the Head and Bill (which is large and strong) to the Tail, decreases in size—The feet are of a lead Colour with black claws, and small in proportion to the size of the Bird. I have seen the feathers on the Head form a more complete Crest. . . ." And in another handwriting there is the additional detail: "Native name is Goo-ge-ne-gang. Likewise it is called the Laughing Jack Ass." Its harsh call must have been one of the outrageous surprises of the Australian bush.

The kookaburra is the largest of Australia's kingfishers. It makes its nest in a hollow in a tree or in a termites' nest. It is a good catcher of reptiles. Diving swiftly to seize a snake or lizard in its beak, it flies high and drops it, diving again before the reptile can escape and worrying it until it is dead. It is also a nuisance to small birds, who know it as their enemy and will sometimes join in numbers to return the attack. An assertive bird, it is always willing to be fed with meat scraps, even in the suburbs.

117

118

The PHEASANT COUCAL was known to the early settlers as a pheasant, no doubt, because of its long, elegantly barred tail. It is, however, a cuckoo, though the only member of this family in Australia that does not deposit its eggs in other birds' nests. Its nest, a globular structure of twigs and grass, is built in a tussock. The coucal lives in the swamps and grassy areas of eastern and northern Australia. Its plumage varies from black to a mixture of brown, black, rufous and yellow, according to sex and to area. It feeds on reptiles, frogs and insects, and is a destroyer of pests such as the grasshopper.

119. Lyrebird. Coloured engraving. Major-General
T. Davies. *Linnean Society Transactions*, 1802.

The LYREBIRD (119–122) was not discovered until the Europeans began to penetrate the
quiet gullies and rocky foothills of the dividing range beyond the settlement. It was found at
the same time as the koala and wombat, by the exploring party in search of a rumoured "land
of plenty" to the south-west of Sydney. Governor Hunter had instructed one member of the
party, his young servant John Price, to keep a journal and on 26th January 1798 Price wrote:
"Here I shot a bird about the size of a pheasant, but the tail of which very much resembles a
peacock, with two large long feathers which are white, orange and lead coloured, and black at
the ends; its body twixt a brown and a green; brown under its neck and black upon its head;
black legs and very long claws." A prosaic description indeed. It was left to later observers to
compare the tail to "the harmonious lyre of the Greeks".

Governor Hunter sent early specimens of the bird, a male and a female, to Sir Joseph
Banks. "They appear to me to be a species of Bird of Paradise," wrote Hunter in the accom-
panying letter. The beautiful tail of the male bird was no doubt the reason for one being
sent as a gift to Lady Mary Howe at about the same time. The first drawing of a lyrebird was
made of this specimen by Major-General Thomas Davies, and an engraving of it (119),
dated 1799, accompanied the publication (1802) of his paper to the Linnean Society; in this
the bird was named *Menura superba*, a reference to its elegant tail. In his *Account of the English
Colony* David Collins described how the two principal feathers formed a frame within which
lay "two other feathers of equal length and of a blueish or lead colour . . . very narrow and
having fibres only on one side of the stem. Many other feathers lay within those again, which
were of a pale greyish colour, and of the most delicate texture, resembling more the skeleton of
a feather than a perfect one."

The lyre shape is more clearly visible when the tail is held upright, but it is more often
carried horizontally to enable the bird to make its way through the scrub. During display the
tail is spread forward gracefully over the bird's back.

Collins also described their habits: "They frequent retired and inaccessible parts of the

interior; and have been seen to run remarkably fast, but their tails are so cumbersome that they cannot fly in a direct line. They sing for two hours in the morning, beginning from the time when they quit the valley until they attain the summit of the hill; when they scrape together a small hillock on which they stand, with their tails spread over them, imitating successively the note of every known bird in the country. They then return to the valley."

This information was gained by Collins, no doubt, from Governor Hunter, who on his part learned it from John Wilson. Wilson, an ex-convict and "wild white man", had reported the existence of a "native pheasant" in 1797, and he was the leader of the exploratory party (including the youth John Price) which in 1798 obtained the first specimen of the bird. Some English ornithologists, who regarded this "pheasant" as a relative of the common fowl, disputed Collins's statements, but it was proved later that the bird did, in fact, build small mounds, give striking displays, and render powerful songs containing much vocal mimicry.

From an early period the male lyrebird was killed for its tail, so that by 1822 a visiting Frenchman, Lesson, said: ". . . it is becoming rarer every day and I saw only two skins that had been preserved . . . during the whole period of my stay in New South Wales." Not only has the bird itself very little natural protection, but it is also a slow breeder, laying only one egg a season. It achieves a measure of protection for the young one, however, by usually placing the nest in a secluded place (on a cliff or stump) with a commanding view.

For some years now lyrebirds have been strictly protected. Today, though not found in any numbers, they can still be heard and seen in bush areas close to Sydney, Melbourne, and Canberra. Their haunts can be recognized by the well-raked forest floor which they turn over with their powerful claws in their search for larvae and crustaceans. And when the male birds display, the forest is filled with the sound of their singing and mimicry.

123

124

125

126

WELCOME SWALLOWS (123) are well known because they often make their nests under eaves. The STONE CURLEW (124) is a nocturnal bird. It makes no nest. John Gould, who named BOWER-BIRDS in the 1840s, described the bower (125) as "the most wonderful example of bird architecture yet discovered". It is not a nest but a playground and courting arena, carefully decorated according to taste. The satin bower-bird prefers blue, the spotted bower-bird chooses white shells or bones. The familiar YELLOW ROBIN of eastern Australia (126) decorates and camouflages its nest with tags of bark. CRESTED PIGEONS (127) are found in small flocks over much of inland Australia. They have a rapid flight and the wings produce a curious metallic sound.

127

There are three drawings of a TAWNY FROGMOUTH in the Watling Collection, each by a different hand and each sufficiently unlike the other to mislead John Latham into giving them different names in the absence of specimens. He classed them all as goat-suckers, but called them Strigoid, Great-headed and Gracile. A near relation, the owlet-nightjar, was called the Crested Goat-sucker and, as Surgeon White explained, the name goatsucker "was given to this genus from an idea that prevailed amongst the more ancient naturalists of their sometimes sucking the teats of goats and sheep; a circumstance in itself so wildly improbable that it would scarce deserve to be seriously mentioned." This odd name is no longer used.

Contrary to common belief, the frog-mouth does not usually catch insects in flight; it feeds on mice, scorpions, centipedes and snails. It sees badly in daylight but is pro-tected by its ability to sleep perched in such a way that it appears to grow out of the tree. In this position the frogmouth apparently feels so secure that it seldom reacts unless actually touched, and then only with a threatening movement of its beak.

The PLAINS TURKEY or BUSTARD (131, 132) was first seen by the men of the *Endeavour* while they were on the north-east coast of Australia. In his journal for 23rd May 1770 Cook mentioned how they found "Bustards such as we have in England one of which we killd that weigh'd 17½ pounds which occasioned my giving this place the name of Bustard Bay."

The bustard once inhabited a large part of Australia but is now extinct or very rare in many once-favoured districts. Although a protected species, it is still hunted for "sport" or for food. It is the size of a turkey and has plumage of subdued browns shading into grey and black, with an edging of black feathers across the chest. The bustard looks particularly handsome in the breeding season when, during courtship displays, pairs strut with their heads high, wing- and tail-feathers stiffened, circling one another, almost beak to beak. The call is a deep booming.

Only a few birds have aboriginal names and the BROLGA (133–135) is one of them. It is also known as the native companion. Some claim that this name derives from its often being found close to aboriginal camps, but it is more likely that *companion* refers to its gregarious nature and *native* merely means *of the country*—a common usage of the word to describe things Australian in the early days. A tall bird, grey-plumaged with red on the head, it belongs to the crane family. Large flocks occur in some parts of eastern Australia, especially the north. The species must have been fairly common near Sydney in the early days, for there is a drawing of it in the Watling Collection.

One of the experiences of the Australian bush is to have the good fortune to see brolgas "dancing". This is a performance which seems to be done in formation, when numbers of them, from a few to several hundred, prance backwards and forwards with a slow measured tread and a rhythmic fluttering of wings. The dance may begin spasmodically with a few birds, and then, suddenly, the rest will join in.

This aerial photograph of a swampy plain and lagoon in Australia's Northern Territory shows how densely the MAGPIE-GEESE congregate. So tightly packed were they that as the aeroplane flew near them, and they took flight, only the birds on the outer edge had enough room to gather speed for their take-off. Those on the inside had to wait their turn. Magpie-geese were once found over a large part of the continent but are rarely seen now in south-eastern Australia.

In the early days of settlement the bird was variously called Pied Goose, Black and White Goose, and Semi-palmated Goose—a reference to its uniquely half-webbed feet. An anonymous note, dated January 1794, accompanied a drawing of it in the Watling Collection: "This bird is about the size of our native wild Goose. . . . It is called by us the New South Wales Goose . . . because its manner as well as its taste and flavour resembles that Bird more than any other. . . . They have only lately been observed, and shot, principally on a Pond near the Hawkesburgh River."

By the 1840s it had already disappeared from this area, and in the whole southern part of the continent it was hunted and its breeding and feeding areas were reduced by the draining of swamps and clearing of land. Recently, in northern Australia, the geese have affected rice-growing, but a report by officials of the C.S.I.R.O., made after lengthy investigation, has stated that the birds "will not be a serious problem to the expanding rice industry." It is added that the species may become extinct in parts of the north unless suitable sanctuaries are provided.

138. Banksian cockatoo. Pencil sketch. S. Parkinson, 1770.
British Museum (Natural History).

PARROTS always suggest the exotic. Some early Dutch maps of the world designated the then unexplored area occupied by Australia as the *Land of Parrots*. A white cockatoo with a yellow crest—one of the parrot family—was brought back to Europe from the East Indies by Dutch merchantmen and explorers before the discovery of Australia. Dampier also described it from those which his men brought aboard in the Celebes: "It is as white as milk, and hath a bunch of Feathers on its head like a Crown." The name *cockatoo* derives from the Malayan word *Kakatua*.

Amongst the birds which Banks saw on the east coast of Australia in 1770, he listed "Parrots and Parraquets most beautifull, White and Black Cocotoes. . . ." He undoubtedly took home several different kinds of parrot, including a black cockatoo and the blue-bellied parrot or rainbow lorikeet. The former was the only Australian land bird to be drawn by the artist Sydney Parkinson during the *Endeavour's* voyage. Alas, it is only a faint pencil sketch (138). Called the Banksian cockatoo, this large black bird with scarlet tail-feathers associates in small flocks of fifty or so and is found over much of Australia.

The SULPHUR-CRESTED WHITE COCKATOO (140, 142), a species related to the white cockatoos found in the islands north of Australia, is the best known and most widely distributed of the cockatoos on this continent. On the open plains, flocks of thousands are common. They feed on native seeds and eat the larvae of a destructive beetle, but this does not win them much sympathy from farmers because of the damage they do to the wheat-crops, and their numbers are constantly being reduced by shooting. A great many also are trapped and sold to be cage-birds. In captivity they have been known to reach a great age—110 years being claimed for the famed Cocky Bennett of Tom Ugly's Point, Sydney. It is not uncommon to find flocks of white cockatoos and GALAHS (141) feeding together. Sentinels are posted to give warning of the approach of danger. One alarmed shriek and the whole flock rises, screeching harshly. This habit explains the Australian slang expression *cockatoo*, meaning a look-out or sentinel, particularly the look-out posted at illicit gambling games. In flight, flocks of galahs with their pink breasts and underwings and their grey backs are a spectacular sight. As they wheel, circle and change direction the colour can flash in a moment from grey to rosy pink and back again. The galah was not discovered until after the crossing of the Blue Mountains, but by the 1840s it was already a popular cage-bird. It is also said to live to a great age.

The ROSELLA (139), with its vivid plumage, was a familiar sight in the early days. Shaw and Nodder illustrated it in their *Zoology of New Holland* in 1792. It was at first called the Rose Hill Parrot, then Rosehiller, after the settlement that is now the city of Parramatta. Like many other parrots, it makes its nest in a hollow limb of a tree or a stump.

139

140

141

142

The BUDGERIGAR (143–145), native to Australia, is now a most popular cage-bird in many countries. It breeds well in captivity. The plumage exhibits a range of colours from the yellowish-green of its native state to bright blue, as well as paler hues.

On the great plains of the inland, huge flocks of these little parrots congregate, feeding on grass seed and never travelling very far away from water. On one occasion, while travelling through this sort of country, we stopped the car and became immediately aware of a strange murmuring chatter. At first we were not sure where it was coming from, but on getting out of the car we heard a noise like rolling thunder—alarming because it was unexpected—as, from the long grass all around us, many thousands of budgerigars took flight. They swept low over the ground in a cloud, then high in the air, and finally settled in the trees, transforming them.

The first budgerigar described by settlers was apparently a solitary bird found near Parramatta, far from its usual haunts. It had probably been driven towards the coast by the drought conditions then prevailing. It was illustrated in *The Naturalists' Miscellany*, by Shaw and Nodder, in 1805, and was called an Undulated Parakeet.

The name budgerigar derives from an aboriginal name for the bird. There are many variants.

144

143

145

Two early names for the WEDGE-TAILED
EAGLE (146, 147) were Bold Vulture and
Mountain Eagle. David Collins described
how the first one captured drove its talons
through a man's foot as it lay trussed in the
bottom of a boat. It has an average wing-span
of about 7 feet and is the fourth largest of the
world's eagles. Some farmers, regarding it as
a menace to stock, destroy it at every opportunity; others regard it as useful because it
kills young dingoes, rabbits, and other pests.

The AUSTRALIAN DARTER (148), a large,
elegant, aquatic bird, frequents the quiet
reaches of rivers and estuaries throughout the
continent. It is known as the snake-bird because of the way it sinks its body beneath the
surface of the water when swimming, leaving
only its long snake-like neck and head visible.

148

This young OSPREY (149–150) was not quite able to fly but it already had a wingspan of about five feet. De Vlamingh, the early Dutch visitor to the Western Australian coast, found a great nest made of boughs which was said to be "full three fathoms in circumference". On the eastern side of the continent Joseph Banks found in 1770, another huge nest on a low island of the Barrier Reef, which he described as "built on the ground by I know not what bird, of a most enormous magnitude—it was in circumference 26 feet and in hight 2 feet 8 built of stick . . .". These were undoubtedly the nests of the osprey, great birds of prey which are found all round the coast. They feed on sea-creatures and occasionally small mammals. The birds add to the same nest year after year until it reaches a great size.

149

It was spring when Dampier was at Shark Bay, Western Australia, in 1699, and he described the nests and young of water-birds. Amongst those he mentioned were cormorants but, like other early explorers, he did not say which sort. The PIED CORMORANT (151-154) is one of several species of cormorant found in Australia. It is widely distributed around the coast and is also found along the Murray-Darling river system, but there is perhaps nowhere in Australia where it is more plentiful than at Shark Bay.

Cormorants (shags) are beautiful birds, especially in the breeding season. They nest in communities on rocky terraces, low bushes or mangroves. The nests are usually packed so close together on a steep slope that there is little room for the birds, returning heavily

laden with food for their young, to alight, except on the few strips which are left between the nests. To reach its nest each bird has to pass by others, the owners of which noisily defend their territories. The parents regurgitate the food (152) and the young ones struggle to get at it. The naked, newborn chicks look startlingly reptilian, but their black skins soon become covered with down.

The FAIRY PENGUIN (155–157) nests in burrows and rocky crannies around Australia's southern shores. It is the only penguin to breed on our coasts. During the day the adult penguins leave their young and go into the sea. On their return they wait until the short period between sunset and darkness before waddling ashore. Then, suddenly, they come unseen on a wave, and as it recedes they stand up. Weighed down with food, they slowly climb the beach to their ravenous offspring and feed them by regurgitation.

The SILVER GULL (158, 159), with its neat grey and white plumage and its red legs and beak, is a familiar sight all round the Australian coastline. It is the scavenger of the seashore. During courtship display the birds stretch their bodies and stiffen their wings, cawing loudly, then bend forward, their beaks almost touching. The chick's patchy brown colouring is a good camouflage.

158

159

The JOHNSTONE RIVER CROCODILE (160) is found in inland freshwater billabongs, lagoons and rivers of northern Australia. It lives on small fish, frogs and crayfish. It is shy and inoffensive, and there is no reason why this harmless creature should not receive the same complete protection wherever it is found as it does in Western Australia. Fortunately, its skin is not considered valuable as leather. The female scoops out a hollow in a sandbank and in it lays twelve to twenty-four eggs. She then covers them over. Staying near the nest, she returns periodically to inspect it. After the young ones appear they stay with the mother for a short time. The freshwater crocodile can grow to a length of eight to ten feet.

The much larger estuarine crocodile, which has been known to reach twenty-five feet, is one of the few creatures dangerous to man in Australia.

The DUGONG or SEA-COW (161, 162) is a sea-going mammal like the manatee of other oceans. All the early explorers found it round Australia's shores. Dampier wrote in 1688, "Neither is the sea very Plentifully stored with fish, unless you reckon the Manatee and Turtle as such."

The aborigines hunt it with skill, seeking out the patches of floating grass which indicate its favourite feeding-grounds. Favoured as food by the early settlers, who thought it resembled

pork, this slow-breeding, harmless creature was quickly reduced in numbers and vanished altogether from southern coastal waters. The dugong's large snout is said to derive from an ancestral elephant stock. The male has tusks which are worn down, and the tough hide which varies from bluish-green to brownish-red is often scarred by scraping against rocks. The flippers are quite flexible and used by the female to clasp her offspring in an upright position while suckling it, so that its head is kept out of the water. Her nipples are visible under her flippers (162). It was probably this strange behaviour which, when first witnessed by incredulous travellers in tropic seas, gave rise to the mermaid myth.

Many GREEN TURTLES were caught by Cook's party at the Endeavour River. Turtles are the longest-lived of all reptiles; a weight of 250 pounds is not uncommon, and 850 pounds has been recorded. They come up the beach to nest several times a year, laying about fifty eggs a time. But lizards—and men too—eat the eggs. The young on their perilous journey (163, 164) down to the water run the gauntlet of birds and lizards, then must face the hazards of the sea. The adult turtle is hunted for the sake of its flesh.

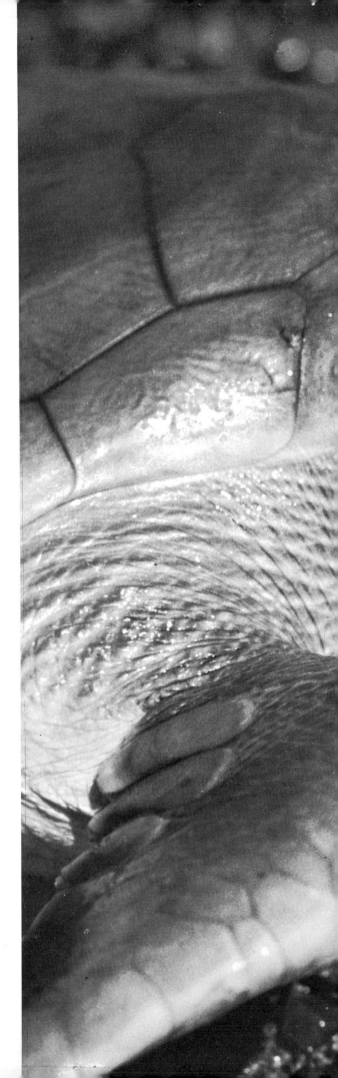

166. Jew lizard. Water-colour. An anonymous artist, Watling Collection, 1790s. *British Museum (Natural History)*.

In Australia there are about three hundred different kinds of LIZARDS. They can be found in a wide range of environments and are remarkably diverse in their appearance and habits. It is impressive to note how they have adapted themselves to living in rain-forests, inland plains, sandy deserts, swamps and rivers. Lizards vary in length from a couple of inches to seven or eight feet. Some can walk on the ceiling with their adhesive discs; others have strong claws. There are those that bark—a small bark—and some can, in an emergency, shed their tails. Many can change colour. There are some with only two legs and others apparently with none. Some lizards bear live young, others lay eggs which are generally buried in the earth. Apart from man, they have many natural enemies including other reptiles, the fox, the domestic cat, and especially birds. Their food is mainly insects, and man should value them for this reason alone. Moreover, none is poisonous and all are completely harmless to humans.

The courtship displays of some lizards, especially monitors and dragons, are spectacular and colourful. Eric Worrell, in his *Reptiles of Australia* described how "the sexes usually move quickly around, darting their heads up and down and lifting their forelegs in a queer circular motion. This appears to serve the purpose of drawing attention to one another. At the same time the male swells the gular sac [throat] and may go through a variety of colour changes". He then seizes the female by the back of the neck.

The BEARDED DRAGON or Jew Lizard was one of the first lizards noticed in the early days of settlement. Several drawings of it by an anonymous artist are to be found in the Watling Collection (166), one of which is accompanied by the note: "This curious lizard (the second yet seen in New South Wales) . . . has a large bag or pouch under the lower jaw which it inflates and contracts at will—when puffed out its appearance is truly singular sometimes resembling the face of a bearded Jew—from which it has here obtained the appelation of the Jew Lizard." This posture is taken up when the animal feels threatened, for its tactics are to bluff rather than to fight. The "frill" is also exhibited in courtship display. The female digs a hole in the ground in which she buries herself and there she lays eight to twenty-four eggs. Emerging from the hole she carefully covers it. Three months later the young hatch out and struggle up to the surface. Bearded dragons eat small snakes as well as insects.

The largest group of lizards, the skinks, are characterized by their smooth scales. The BANDED SKINK (168), pale brown with darker brown crossbands, has a snake-like appearance although it has legs. It lives amongst rotted timber and leaf mould. The young are born alive.

GOANNAS or MONITORS (169-175) are the largest Australian lizards and the PERENTY (or perentie) is the biggest of these. This perenty (169) from the plains of northern Australia was exceptionally long, nearly eight feet. It had probably fed well on the young water fowl of the lagoons. When surprised it raised itself high on its legs, arching its back, hissing, and darting its bifurcated tongue in and out incessantly. Sometimes goannas will stand up on their back legs only, to get a better view or to fight with each other. The large powerful claws are useful in tree-climbing. When alarmed they frequently climb the first available vertical object. Eric Worrell, of the Australian Reptile Park at Gosford, tells of seeing a goanna, when pursued, clamber up a man and perch on his head. Goannas feed on carrion as well as on live prey—such as rabbits, smaller reptiles and insects.

167

168

At close quarters the THORNY DEVIL with its sharp spiny scales looks formidable, but it is quite harmless. At the most it grows to 8 inches. It eats thousands of black ants a day. The young hatch out from eggs buried in the earth. The four young ones seen on the hand in pl. 178 are one day old.

179

180

181

182

Surely Australia's best-known lizard must be the COMMON BLUE-TONGUE (179). Often seen in suburban gardens, it can be recognized by its beautifully shaped blue tongue which it shows whenever it feels itself threatened. Its presence should be welcomed because of the snails and other insect pests it eats. Some people, however, mistake its triangular-shaped head for that of a death adder, and kill it. This lizard, which averages some twenty inches in length, gives birth to live young and can have as many as twenty in one litter.

The SHINGLEBACK is perhaps more commonly known as the bobtail lizard because of its stumpy tail. The scales have the appearance of a pine cone. It is well known all over mainland Australia. The most usual colouring is brown with light markings (180), but many colour variations are known, and some shinglebacks are almost black (181, 182). Of special zoological interest is the evidence of a vestigial pineal eye which is developed in similar fashion to that of the ancient New Zealand lizard, the tuatera.

The SPINY-TAILED SKINK or Gidgee Lizard (183) has a flat, depressed tail bristling with spines. When it takes refuge in rocky crevices, under stones or in cracks of tree-trunks these spines make it almost impossible for an enemy to dislodge it.

This GECKO (184) was an attractive pet. Its body was a beautiful rich velvety brown with yellow markings. Apart from being able to walk upside down on the ceiling, the gecko has the extraordinary habit of being able to shed its tail and grow a new one. If a bird attacks it, it sheds its tail in an instant. The piece jumps around on the ground of its own accord, diverting the bird's attention and allowing the gecko to escape.

185. Snake. Water-colour. Signed Thomas Watling, Watling Collection, 1792-6.
British Museum (Natural History).

"Venomous animals and reptiles are rarely seen. Large snakes beautifully variegated have been killed, but of the effect of their bites we are happily ignorant," wrote Captain Tench in 1788.

Much the same may be claimed today, for although Australia has about 140 different kinds of SNAKES, about twenty of which are dangerous to man, the average Australian rarely sees them. They are there all the same for those who know their haunts and their habits. The colour and markings of most provide camouflage and they are quick to glide away if disturbed. Even the more venomous ones seldom behave aggressively towards a human being, and if a snake strikes it is probably because it has been surprised, cornered or injured. The best known of snakes dangerous to man are the common brown snake, the common black snake (186, 188), the copperhead, the tiger snake (187, 189–93) and latterly the taipan.

These photographs were taken on a tiger-snake-collecting trip with Eric Worrell of the Australian Reptile Park. He catches the snakes and extracts the venom which is used by the Commonwealth Serum Laboratories to make an anti-venene. This antidote has saved many lives since it became available, for the venom of a tiger snake is, weight for weight, one of the most deadly poisons in the world.

Tiger snakes are not usually more than $4\frac{1}{2}$ feet long. They can vary quite a lot in colour but are generally grey-brown with yellowish bands. They live in swampy areas where there are plenty of frogs and mice to feed on. The tiger snake in pl. 187 had just killed a frog which it was swallowing head first. As a snake cannot chew or tear its prey it must swallow it whole. The flexible hinging of its jaws enables it to open its mouth wide enough to admit an animal larger than its own head. It uses its backward-pointing teeth to work the prey gradually down its gullet and into the stomach where it is digested.

(188)

When a snake strikes (190-193) it sometimes twists its body into an S-shape and, raising its flattened head, throws itself at its prey.

During courtship the male caresses the female by gliding back and forth along her body (188, black snakes, and 189, tiger snakes).

Some snakes lay eggs, others give birth to live young. The common black snake usually bears about twelve live offspring, but the tiger snake can produce as many as fifty. Once they are born the young ones must fend entirely for themselves.

(189)

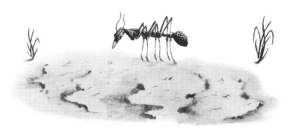

"Of insects here were but few sorts and among them only the Ants were troublesome to us." This was the opinion Banks formed in his summary of the discoveries in natural history made by the party on the east coast. He went on to describe in particular the GREEN ANT (195-197) as being "green as a leaf and living upon trees where he built his nest, in size between that of a mans head and his fist, by bending the leaves together and glueing them with a whiteish papery substance which held them firmly together. In doing this their man[a]gement was most curious: they bend down leaves broader than a mans hand and place them in such a direction as they chose, in doing of which a much larger force is necessary than these animals seem capable of. Many thousands indeed are employd in the joint work; I have seen them holding down such a leaf, as many as could stand by one another each drawing down with all his might while others within were employd to fasten the glue. How they had bent it down I had not an opportunity of seeing, but that it was held down by main strengh I easily provd by disturbing a part of them, on which the leaf bursting from the rest returnd to its natural situation and I had an opportunity to try with my finger the strengh that these little animals must have used to get it down. But industrious as they are their courage if possible excells their industry; if we accidentaly shook the branches on which such nest[s] were hung thousands would immediately throw themselves down, many of which falling upon us made us sensible of their stings and revengefull dispositions. . . ."

The glue is in fact silk, which is squeezed from green ant larvae that have been carried to the leaf-edge by worker ants, which, using them as shuttles, weave the leaves together.

After experiencing the bites inflicted by green ants, Banks was not inclined to think TERMITES (198-200), otherwise known as white ants, much bother. "We have ever since we have been here observd the nests of a kind of Ants much like the White ants in the East indies but to us perfectly harmless; they were always pryamidical, from a few inches to 6 feet in hight. . . . Today we met with a large number of them of all sizes rangd in a small open place which had a very pretty effect; Dr Solander compared them to the Rune Stones on the plains of Upsal in Sweden, myself to all the smaller Druidical monuments I had seen."

There are many species of termites and the population of colonies can range from less than a hundred to two million. Their mounds of cemented earth contain a maze of tunnels and galleries, housing the royal cell where the termite queen is kept, the nursery where the eggs and young are tended, and the galleries where the food is stored. A mound twenty-four feet high has been recorded. Sometimes if a mound is knocked over, the termites continue to build upwards (200).

The destructiveness of termites is one of the great problems of northern Australia. They will attack all kinds of woodwork as well as growing trees and plants, and have been known to destroy the lead sheathing of underground cables.

ILLUSTRATIONS